DREAMS UNDER CONSTRUCTION

"Dream big and know that all things are possible."

— John Bahen

DREAMS UNDER

CONSTRUCTION

THE LIFE, WORK AND LEGACY OF JOHN BAHEN

Eloise Lewis, Editor

Copyright © 2014 and 2017

All rights reserved. No part of this publication may be reproduced or transmitted in any form or by any means, electronic or mechanical, including photocopying, recording, or any information storage and retrieval system, without permission in writing from the publisher.

First edition published privately in 2014 by
ISBN 978-1-927483-84-8
BPS Books
Toronto and New York
www.bpsbooks.com
A division of Bastian Publishing Services Ltd.

Second edition published and distributed by Kinetics Design – KD Books in 2017
ISBN 978-1-988360-06-5
www.kdbooks.ca, www.linkedin.com/in/kdbooks

Edited by: Eloise Lewis, LifeTales, www.lifetales.ca

Cover, text design and typesetting: Daniel Crack, Kinetics Design – KD Books

CONTENTS

ACKNOWLEDGMENTS xi

1 OPENING DAY 1
2 GROWING UP BAHEN 9
3 ENGINEERS ARE BORN, NOT MADE 17
4 BUILDING A LIFE OF BUILDING 23
5 SOUTH TO THE TURNPIKE 32
6 NORTH TO GOOSE 39
7 LEARNING ON THE JOB 47
8 BIDDING FOR AMBITION'S SAKE 55
9 KIEWIT ENCOUNTERS 63
10 WEIGHING HIS OPTIONS 68

11	THE WELLAND	73
12	PROJECT OF THE CENTURY	78
13	FROM ROADS TO DAMS AT JAMES BAY	85
14	BUMPS IN THE ROAD	93
15	THE BURDEN OF JUST ONE ROCK	101
16	A FULL PLATE	110
17	BAHEN-STYLE	117
18	INTO THE WEST	131
19	PLANNING THE NEXT MOVE	144
20	TIME TO GIVE BACK	151
21	PUBLIC RECOGNITION	159

22 LIFE CHANGES 165

APPENDIX 1 175
SPEECH BY JOHN BAHEN AT THE OPENING OF
THE BAHEN CENTRE FOR INFORMATION TECHNOLOGY,
UNIVERSITY OF TORONTO, OCTOBER 8, 2002

APPENDIX 2 179
JOHN BAHEN CITATION UPON HIS BEING GRANTED
AN HONORARY DOCTORATE OF ENGINEERING DEGREE,
UNIVERSITY OF TORONTO, JUNE 9, 1999

APPENDIX 3 183
AN ADDRESS PRESENTED TO THE UNIVERSITY OF
TORONTO GRADUATING CLASS OF ENGINEERS RECEIVING THEIR
BACHELOR OF APPLIED SCIENCE DEGREE, JUNE 9, 1999

REMEMBERING MARGARET AND JOHN BAHEN 189

ACKNOWLEDGMENTS

THIS book has been proudly produced with the help of many contributors. John Bahen himself was initially closely involved with the storytelling, as were many of his family members, friends and former colleagues. Copious interviews were conducted, and technical questions asked, in order to recount John's brilliant engineering career and capture glimpses of his family life. The Bahen family wishes to thank Patti Hall, who wrote the original manuscript, based on her excellent research, interviewing and writing skills.

Special thanks to Gene Bednarski, Jocelyne Desrochers, Denis Evans, Geoffrey Lindup, Donald Quane, Walter Scott, Jr., Colin West and John Wilkes for their perspectives on John, both professional and personal; to Robert Prichard and Michael Charles for their input on John as a philanthropist and University of Toronto alumnus; and to his family – Margaret, Stuart and Susan, and Ted Chant – for sharing their stories about John, the family man.

Eloise Lewis
Editor

*John speaking at the opening of the Bahen Centre for Information Technology, University of Toronto.
The platform party at the opening of the Bahen Centre for Information Technology, U of T.
(Photos: Steven Evans)*

1

OPENING DAY

JOHN BAHEN looked out from the podium, smiling broadly at the invited guests and acknowledging their warm reaction to his introduction. He was accustomed to the formality of such an occasion and comfortable in front of a crowd. An experienced speechmaker, he conveyed the essence of his character with his opening line.

"Clearly I am speaking from the bottom of the batting order on the list of speakers. For those of you who know me, this is a position I enjoy occupying. I am used to having the last word!"

It was October 8, 2002, the day of the official opening of the University of Toronto's (U of T) Bahen Centre for Information Technology. The atrium of the building was awash in sunlight. On a platform backed by the restored brick wall of the abutting Koffler Centre, tall banners listed the sponsors, donors and quasi-governmental bodies responsible for funding construction of the new Bahen Building. A banner bearing a portrait of John and his wife Margaret read: "Margaret & John Bahen – Leaders, Builders & Graduates." A blue satin ribbon, tied in a large bow, was strung across the width of the platform in front of a podium blazoned with the university crest. The platform itself was adorned with the flags of Canada, Ontario and the University of Toronto.

The conditions of this event were special to John in ways not self-evident to many of the people gathered or passing by. Many students paused on their way to classes to observe the proceedings. They stood in the main hall, leaned over balcony railings and peered out of crowded lecture hall doorways. Few of these engineering hopefuls likely realized that the man speaking was one of Canada's most successful construction engineers. Fewer still likely knew that he was an esteemed U of T alumnus, co-chair of the highest-yielding fundraising campaign the university's Faculty of Applied Science and Engineering ever had, winner of the Engineering Alumni Medal, recipient of an Honorary Doctorate of Engineering and a Member of the Order of Canada.

Here was a man who had worked on some of Canada's largest infrastructure construction projects in the latter half of the twentieth century, including Highway 401, the Welland Canal and St. Lawrence Seaway in Ontario, the James Bay Hydroelectric Project in Quebec, port developments on both the Atlantic and Pacific coasts, main-line irrigation projects in Alberta, the Nipawin Dam in Saskatchewan and several contracts for Vancouver's elevated rail transit system. Nor would they have known that he was involved in the construction of the subway that brought many of them downtown to school and of the system of expressways and highways that others drove to travel home.

Two men shared the platform with John. The dean of Engineering, Michael Charles, stood on one side with speaking notes in hand. On John's other side was his caregiver, standing close enough to discreetly offer the support he occasionally needed, in the wake of a stroke a year earlier.

John spoke slowly, in short sentences, pausing in a practiced way between phrases, allowing time for the polite chuckles that greet light humour in otherwise formal speeches. Looking out at the audience over the rim of his reading glasses, he was clearly enjoying their reactions to his self-deprecating jests and political quips. Margaret, sitting in the front row of the audience, at first listened nervously, realizing that her husband had diverted from the speech they had practiced together and was now ad-libbing at will. Her nerves quickly calmed as it became evident that he was performing magnificently and relishing every minute of it. (For the full text of the prepared speech, see Appendix 2.)

This speech was personal for him, and he spoke passionately. From the time he was a boy, John had wanted to build. And build he did – a dream shared by many in attendance as well as those passing by. Declaring the Bahen Centre for Information Technology officially open for classes was the culmination of many years of advisory and generous philanthropic work for John, a powerful believer in maintaining U of T's status as one of the finest engineering schools in the world. For the same reasons that he supported his alma mater, he was committed to fuelling the future of young engineers. He remembered the days he had been an ambitious engineering student himself and had, from time to time, struggled with his course load. Now a major faculty building bears his name. His career had witnessed five decades of technological change and he was successful in advance of today's modern tools.

John's advice to all engineering students was to practice what he often says cannot be taught: "Dream big, and know that all things are possible."

The name John Bahen is not well known outside the Canadian construction industry. The people behind the scenes of massive construction projects and similar undertakings rarely are. Yet, fifty years after graduating from U of T, he stood at the podium to express his gratitude for the university's crucial role in his success – the success that made possible his lead financial donation to that building's construction fund.

John acknowledged that the much-needed building would not have been possible without federal and provincial government funding support, but he also reminded listeners that private sector donors, many of whom were proud alumni, raised some $100 million. He told the audience, "The future of our country will likely walk the halls of this building someday – maybe even tomorrow. That makes me very proud."

John emphasized the undeniable impact that information technology has had on all economic sectors. He entered U of T as a civil engineering student at a period when the Faculty of Applied Science and Engineering faced a space limitation due to unusually high enrolment,

DREAMS UNDER CONSTRUCTION

one that required facility expansion. It was during this post-war period, from 1945 through 1949, that engineering programs across Canada were deluged with applications from veterans and demobilized soldiers. He knew applicants who had been turned away and was familiar with the extreme measures often necessary to accommodate as many as possible of the exceptionally qualified students rather than lose them to other schools or professions.

John never understood why the university was unable to anticipate the enrolment surge at the end of World War II. His disappointment at the insufficient planning of the late 1940s convinced him to assist in every way that he could in the late 1990s, so other classes of engineering applicants would not be turned away because of space shortage and professor scarcity. At that time, there was a flood of applicants, caused by the "double cohort" graduation from Ontario's secondary schools when the government eliminated grade 13. The Bahen Centre for Information Technology is a key part of the university's response to this surge in applications.

As a few hundred guests mingled following the ceremony, conversations were broken by the sounds of the brass band that is one of the many claims to fame of the U of T Engineers – the Lady Godiva Memorial Bnad – playing the Engineer's Hymn. Typically, in a setting where professors and administrators are robed and hooded and guests dressed formally, interruption by a group of students in hockey jerseys, wearing yellow or red decorated hard hats, would be considered an event crasher. But this was one of U of T's homes to engineering, and the "Bnad" (and other infamous elements of "Skule" spirit) is its heart.*

---

* The Lady Godiva Memorial Bnad is a brass band run largely by engineering students at the University of Toronto. It made its debut in 1950 on a flatbed truck during a homecoming parade. The band, together with other traditions including Ye Olde Mighty Skule Cannon, which is the engineering school's mascot, is a cornerstone to Skule™ spirit. The Bnad's misspelled name (as the school's misspelled name – Skule) references the resistance or inability on the part of engineers to consider proper spelling. The refrain of the engineers' cheer is "We are we are we are we are we are the engineers / We can we can we can we can demolish forty beers."

~ 4 ~

OPENING DAY

The crowd gladly opened a passageway as the motley troupe, led by a flag bearer, moved slowly toward the stage to the beat of the bass drum. Smiles, laughter and loud applause greeted the students' procession.

Rather than being shooed away, the leader of the band (still holding his saxophone) was ceremoniously guided onto the platform and offered the microphone. After a welcome on behalf of all students and an informal thank you to those who made this day possible, he reminded the crowd that U of T engineers are an entity unto themselves with a proudly defended legacy.

No one enjoyed the break in formal ceremony more than John Bahen.

A narrative on the subject of John Bahen's life is a reminder of the

The Lady Godiva Memorial Bnad.

history of the engineering profession for the last several decades. It is a profession that has evolved in response to societal, environmental and technological change and one that deserves greater awareness, given its pivotal importance in everyday life. As people go safely about their lives, the fact is that virtually everything – including the vehicles they drive, the streets they use, the water from their taps, the power behind the light switch and the buildings and infrastructure where they live, work and play – is the result of the vision, skills and work of engineers. Canadians are not alone in their blindness to engineering. The lack of visibility and resulting lack of recognition of the impact of the engineering profession on our society is common all over the world.

Certainly it would serve us well to remember. One way to appreciate accomplishments and successes is to pay respect and homage to those who have influenced us. Such is done frequently and with considerable ceremony in the fields of politics, medicine and the law. But, in the case of civil engineering, little is known of the individuals, teams and companies that literally build our changing world.

John has been a witness to, as well as a product of, change in the last half century. As such, he is a shining example of resilience, adaptability and innovation. He has been described many ways: a character, the eternal optimist, an indomitable spirit, almost irrationally driven, an unstoppable force, passionate about his work and grounded in the steadfast belief that all problems can be solved.

It isn't luck exactly, for an engineer to find himself at the right place at the most opportune time. Engineers are lauded for planning ahead, designing to exact specifications, leaving nothing to chance and putting countless backups in place for the just-in-cases. John knew that he wanted to build and that an engineering education was his essential first step. He might have been absorbed into the large graduating classes filling a great number of entry-level jobs, the majority of which were still managerial, especially in the 1940s. He might have been unable to easily distinguish himself from the countless other civil engineering grads who wanted jobs close to home, with responsibility and great salaries. But distinguish himself, he did.

John in 1939.

2

GROWING UP BAHEN

IN February 1927, John's parents, George and Katherine (Kay), boarded the RMS *Ausonia* in Southampton, England, with their seventeen-month-old toddler William to travel to Halifax, Canada. George, a Canadian citizen, had married Kay, who was English, while he was working in England after the war. Their second son, John, entered their lives later that year, on December 27.

George was a second-generation Canadian. His paternal grandparents, the Behans ("Behan" was the original Irish spelling of the surname), were Roman Catholics who had immigrated to Carleton County, near Ottawa, where George's father, Thomas, raised his family.

John's grandfather, Thomas Bahen (by then the name had been Canadianized and Protestantized), was listed as a labourer in one census and, years later, in 1894, at the time of George's birth, more specifically as a broom maker.

George served in England during World War I. On a 1920 ocean passage declaration, George's intended occupation was listed as "chemist." Although travel records show that he came home after the war to visit his family in Guelph, Ontario, he returned to England for some years. There he met and married Katherine Simmonds, a resident of Middlesex. Three years later, after landing in Halifax, they began their lives in Toronto, Ontario.

Just a few years later, George went to New York State for work, where he died, apparently a victim of pneumonia, penicillin not yet being available. Kay found herself the single parent of two young boys in a still unfamiliar country. Although it was not uncommon then for a woman to be the only parent of a household, few support services were available. In the face of what must have been a daunting outlook, she was forced to make challenging choices. Finding herself in need of childcare in order to work the long hours required to support herself and the boys, Kay arranged for Bill and John to live at Earlscourt Children's Home. It must have been the most difficult of the choices she faced as a young mother.

Earlscourt was a small church-run facility for children exposed to stressful family environments or part of "broken" families. It was home for the Bahen boys from not long after their father left until Bill reached secondary school age and could be partially responsible for his younger brother after school. The home was operated by the Earlscourt Methodist Church and received funding support from both the church and a local charitable organization. Earlscourt was the vision of its minister Reverend Peter Bryce, a dedicated social activist and community advocate. Opened originally as a daycare centre in a small house on Dufferin Street in Toronto, it became, in 1918, a full residential home located at 46 St. Clair Gardens. During wartime, Earlscourt provided care for the children of serving soldiers whose wives needed to work to supplement the family income.

The home received dedicated financial support from the Wimodausis Club (a name derived from the first letters of **wi**ves, **mo**thers, **dau**ghters and **sis**ters), whose members united to raise funds, primarily through the sale of an annual cookbook and by hosting sales and bazaars. It was common to see the Wimodausis Club cookbooks advertised for sale in the *Toronto Star* in the 1930s and 1940s. An example of one of the earliest charitable women's organizations in Toronto, Wimodausis also donated services in educating women on "the perfectly appointed table" including the arrangement of linen, china, crystal and flowers and assistance in selecting accessories in the Toronto Eaton's store. The same organization raised funds for not-for-profit organizations well into the 1990s from its Toronto Antique Show.

About forty children lived in the home, ate meals, prayed together and attended church and the nearby Earlscourt Public School. Neither of the Bahen brothers recalls being lonely there, thanks to the structured, caring atmosphere of the large, busy house managed by Miss Hattie Inkpen and her staff. The reasons for a child to be accepted at Earlscourt were as numerous as the children themselves. The death of parents, family breakup, financial hardship and being born to an unwed mother are some of the reasons provided in historical accounts. John does not remember such private matters being spoken of. An atmosphere of acceptance prevailed.

"That was just the way it was and it wasn't appropriate to question. I was just a little kid with glasses. We never thought to ask why. We carried on."

John remembers the structured atmosphere of the Earlscourt Home with gratitude and a warm sense of nostalgia. The other resident boys became, over time, more like brothers than classmates and he fondly remembers them by their nicknames. There was Gummy, Scotty, Dougy and Sargey, while John was known as Baheny. Dougy Banfield was John's best friend and sidekick, and, for five years of their childhood, the two boys slept in adjacent rooms. They remained close, both joining Sea Cadets after they returned to their own homes and attended high school.

John finds nothing but value in having resided in what today would be considered an orphanage. It was more than a surrogate home; it was the setting for his dreams and single-minded ambition. As the saying goes, engineers are born, not made.

"At Earlscourt," John recalls, "that's where it all began. They had a great big sandbox, and I took it over and built my city in it with highways and parks. I had little cars, toy cars. But no lakes or dams yet!"

Every Saturday, after chores were done, Dougy and Baheny

1. John with his older brother, Bill.
2. John with Nora Noble.
3. John the Sea Cadet.
4. John with his mother, Kay.

commandeered the sandbox. Baheny was in charge; he had the vision – he was the engineer, the planner and the builder. It was his city, but without Dougy there would have been no way to keep other kids from touching that sandbox, altering the plan or destroying their ongoing work.

Even then, John's need to build things was more powerful than any other, though school and chores had to come first. For pocket money, John sold newspapers on the street with his brother and their friends, the Heisey brothers. He was active in sports, and outside whenever he could be, but there always had to be a project underway. The sandbox build was not just entertainment and the means of solidifying his friendship with Dougy. It was part of John's makeup that he coped with any difficulties by channeling his energies into creating and building.

In the sandbox, the roads were two cars wide, making transportation corridors. The cars, which they formally called their "vehicles" (this was serious engineering business), were carefully placed on the roads, which had been swept into a smooth surface by the back of a chalkboard brush. The city had places for living and working and places for playing baseball, and its terrain was never even (which made the driving more fun). Baheny had thought about these things.

"Highway systems, bridges, subways and all the big heavy construction stuff like dams and powerhouses – that's what I've been," he says. "I graduated myself an engineer right out of the sandbox."

The most enduring feature of the Earlscourt years was the summer camp on Lake Simcoe at Jackson's Point administered by the Methodist Church. Imagine the relief of city-bound children who were housed relatively tightly in one large building less than a block from their school when released to the natural freedom and relaxed atmosphere of summer camp for two months. The children were supervised at camp as conscientiously as in the city. Their summers were much like those of children who had a more traditional full-time family-based home life.

Their strongest memories are of the lake, swimming, bonfires, games, beachfront building, competitions and sleeping outdoors.

John and his friends found their leader in Bill "Sargey" Sargentson, a boy who spent years at Earlscourt as a resident and then assumed a counsellor role working with the younger children after he returned home. Sargey also happened to be a very close friend of John's older brother Bill, a friendship that lasted throughout the men's lives. Sargey had a crush on a local girl, Nora Noble, who spent her volunteer summer hours helping out at the Jackson's Point camp. Nora's family was well respected in the small town of Sutton; in fact, her father was the town doctor. The Noble family farm and home, which housed Dr. Noble's office, was a cornerstone of the community.

Thanks to Nora, the Bahen boys became very close to the Nobles and worked on the farm throughout the year whenever they were needed. Once the Noble family sons both went to war, Dr. Noble did as much as he could on the farm, when he could spare time from his medical practice, but he needed help to maintain it. In exchange for their assistance, the Bahen boys, with their mother's blessing, spent a great deal of time in the company of the Nobles, where they were treated like family.

Dr. Noble's busy household was frequented by his many well-educated and successful professional friends, including writer and humourist Stephen Leacock. This provided a far greater life experience than would have been possible for a pair of boys living in an urban home for children in need. It was an extended family of sorts for the Bahen boys, who had never known a family environment. The Nobles acted as mentors to young John and encouraged him in his education. As a result, university became an expected goal.

While Bill and John were at Earlscourt, their mother was employed at Eaton's Round Room restaurant, on the fifth floor of the College Park store, and lived in a humble one-bedroom apartment. Later on, when the boys were in their teens, she took the position of manager at Childs

Restaurant on King Street, the well-known financial district eatery. Kay's growing network of influential Toronto customers had a marked impact on John's life. Many important business leaders lunched at Childs and reserved the same table each day through Kay. When John visited his mother at work, he met these patrons, some of whom offered him odd jobs or summer work. Many became cherished family friends, including Timothy Eaton, Jr. John was agreeable and outgoing, and his willingness to try any kind of work brought many early mentors into his life.

John credits his delicate, English-accented mother with both his commitment to achieve the highest possible level of education and his sense of humour. Kay worked long hours in the restaurant every day but Sunday, and John believes that these physical demands, combined with her perfectionist personality, affected her health. She was stricken by a series of serious illnesses and eventually died of heart problems, in 1959, when she was only in her fifties.

John (left) walking with a friend on King Street; Childs Restaurant, where John's mother worked, is in the background.

John (left) with his good friend Bob Bertram (right) and another fellow student in front of House 731.

3

ENGINEERS ARE BORN, NOT MADE

By 1947, not only did John know he would go to university, but he also knew it would be to study engineering. There was "no other choice that made sense" than to enrol in the nearby University of Toronto's engineering program. His older brother was already a student at U of T, pursuing a law degree. The fact that the university was close to home allowed the boys to be a support to their mother. The more powerful influence on John's choice, one that was predictive of the importance he would later place on being a mentor himself, was the advice of two prominent figures in his life, Col. W.S. (Stewart) Wilson and William "Bill" Tate. They told him that if he wanted to be part of a team that built big impressive structures, the best foundation was a degree in civil engineering.

In the summers before high school graduation, the business-minded John gathered boys from the neighbourhood or fellow students of Vaughan Road Collegiate to make money cutting lawns and doing gardening. They used a half-ton truck and equipment belonging to one of the older Vaughan students, Bill Canning, and did whatever yard maintenance work they could find. One of the first lawns that John cared for, in the summer of 1946, was that of the Wilson family on Humewood Drive. Anne Wilson, one of the family's three daughters, was then dating John's brother Bill.

John came to know Stewart and Eleanor Wilson as closely as if they were his own parents. Their home became another welcoming family environment for him. As luck would have it, when John and his buddies took over the lawn care, Stewart Wilson was the Assistant Dean and Secretary of the Faculty of Applied Science and Engineering (FASE) at U of T. When it became evident to Wilson that John possessed the work ethic and natural mechanical skill to be an engineer, he strongly encouraged him to apply to the civil engineering program at U of T – and apply early. Although John never knew for sure if Wilson had a reason to remind him repeatedly to get his application in early, it may have helped him get one of the positions held open for civilian entrants at a time when a large number of returning veterans were being given priority for spaces.

The other influential mentor for John, Bill Tate, lived near the Wilsons. Tate was both a close friend of Colonel Wilson and, coincidentally, a regular lunch customer at Childs Restaurant when Kay Bahen was running it. John's education choices and, by extension, career, may tie directly back to Tate, who was a 1909 alumnus of the Civil Engineering Program at the University of Toronto. At the time they met, he was General Manager of the Toronto Transit Commission (TTC) and extremely well known in the city where the subway construction project was drawing North American attention.

Tate and Wilson were fishing buddies as well as neighbours. Both properties were John's regular yard work responsibility and his work occasionally extended to cottage repair and other small mechanical jobs, such as helping when the two men needed to induce a small outboard motor into operation. The obstinate motor had been stored for the winter, and neither had been able to get the thing going. One day, in 1946, they had John give it a try in a forty-five-gallon drum of water. As if by magic, he got the motor started. John became a regular invitee on short fishing trips to Lake Simcoe and the Severn River, but he was under no illusions – he was there to keep the boat running. John had begun the steep climb toward his dream by grade 13, not knowing of any program or institution that might better fit his needs. The admission period that today occurs over six months took only a few weeks then. It was difficult to get all of the paperwork to the admissions office in time for a decision to be made. Often students were going in on the first day of classes to find out if they had been accepted. Also, unlike nowadays, students needed to prove that their final secondary school year included credits in English and history and four other credits in the course load. Although a math credit was encouraged, there was no requirement to have taken a senior level math for admission into engineering. This is unthinkable to engineers who studied after 1950 and certainly to those who struggle to obtain high standing in three university-level math courses to get into engineering programs today.

Applicants were also asked to supply a Certificate of Good Character, more a personal reference than a testimony to specific skills present in successful engineers. William Tate offered the reference for John. But the relaxed technical standards of pre-World War II engineering were a result of the fact that the profession was simply more managerial then. Soon after war's end, technical advancement in popular engineering fields required significant upgrades in admission standards and course content. In fact, grade standards were not applied until 1950-1951, when a 60 per cent average became the minimum.

In his opening comments to John's first-year class, Dean Young said that the advent of technologies requiring highly skilled engineers "of the hands-on variety" would demand a curriculum upgrade. That upgrade at U of T and every other noteworthy engineering school was perfect timing for John as he began his engineering studies.

Despite his busy summer yard work business and short-term jobs such as waiting tables at Fran's Restaurant and assembling toys and bikes at Eaton's, the tuition fees and costs of attending university were out of his immediate reach. John and his brother were, after all, the children of a single, hard-working parent. Unwilling to put additional pressure on his mother, and without other options, the ever-optimistic John walked into the King Street branch of the Canadian Imperial Bank of Commerce and applied for a loan. This was not the first time that unbridled confidence would pay off for him. The round-faced manager of the bank, also a daily lunch customer at the restaurant his mother managed, had always waved John in to his office and asked if the bank could do anything to help him. For more than a decade after graduation, John remained a customer of this manager, who had approved him for a $400 loan as a first-year student. That amount was just enough to cover his basic expenses, including the $300 academic fee.

Acceptance letter in hand and tuition fees paid, John was notified that, instead of attending classes and participating in campus life on the picturesque St. George campus, he would be attending classes in the Ajax campus portion of the freshman class, twenty-five miles away. He was pleased about this, afraid that the downtown campus would be a distraction for him, one that might compromise his good grades and make a speedy completion impossible. School was only a means to an end for him – a way of stacking the proverbial deck for access to great positions and challenging projects. As it turned out, John took quite a long time to complete his degree, but what drew his attention away was the temptation to work on distant job sites. He was distracted, indeed, and when he found himself dismayed with his grades, the desire to build won out over staying in school.

Undergraduate enrolment had begun to increase at the University of Toronto's Faculty of Applied Science and Engineering, which, in 1936, had over a thousand students. Steady growth, owing to increasing interest in the program's offerings and job opportunities in engineering, combined with the post-war veteran education policy of financial support for post-secondary or retraining programs, precipitated an unprecedented doubling of applications by 1946.

With an existing capacity for only three hundred fifty first-year students on the main St. George campus, and projections of up to fifteen hundred applicants for the 1945 fall term, U of T leased approximately three thousand acres of land that comprised the site of Defence Industries Limited (DIL), considered the largest munitions plant in the Commonwealth. The Ajax campus used 111 buildings of the former munitions plant and took advantage of previously existing services such as a hospital, gymnasium, theatre, dance hall, sports facilities (hockey, tennis and baseball) and a massive cafeteria. Some fourteen hundred students commenced studies at Ajax, 80 per cent of them veterans, with another four hundred first-year students attending class downtown.

By September 1947, when John entered the engineering program, over four thousand students were enrolled across the two campuses, a level that, while not sustained for long, represented the greatest pressure ever on the faculty to provide facilities and qualified instructors. By comparison, some fourteen hundred undergraduate students are currently enrolled in the faculty. Even the double cohort graduation in Ontario in the late 1990s, mentioned for its role in prompting facility expansion and an increase in the number of endowed teaching positions, paled in comparison with the post-war application bulge.

"A-jacks," as they were known, attended classes at the Ajax Division between 1946 and 1949. Accounts from those years show that the students held a deep affection for Ajax, and John was no exception in this. Stories usually mention roommates and housemates, the buses, dances and parties, the campus dog (Big Red) and the food (generally unappreciated, although John had no complaints). The grounds of the former munitions plant were unusual, with overhead pipes and boardwalk sidewalks. In all, Ajax was a unique campus experience. Despite the intensity of its academic regimen, there was a very active social life on campus for those in residences. The extended familial atmosphere that seems to have characterized Ajax was a function of the high priority placed on the social aspect by the students themselves. One reason was that the academic demands were so extreme, and every student there was in the same situation. The interesting campus and the social network were instrumental in keeping John in school.

John's roommate in his first year, Bob Bertram, convinced him to manage his campaign for House Master so they could enjoy the social benefits of having the largest room in House 731. After the successful election, the two occupied the corner room and, as they maintained an open-door policy, never wanted for company. Among other tasks essential to a collegial atmosphere at Ajax was that of organizing house events. This ran the gamut from convening parties at a local golf club to recruiting female employees from Bell Canada's service centre in downtown Toronto to attend regular Friday and Saturday dances and house parties. John sums it up this way: "It was a blast at Ajax – completely different than downtown."

It was nearly Christmas 1948, during John's second year, when bad news turned his world upside down. His older brother, Bill, was travelling south with Anne Wilson from an event in Owen Sound when an oil truck lost control on the highway, hitting their car. Anne succumbed to her injuries in hospital. Bill, at that time still a law student at U of T, spent months in hospital.

The months of hospital care, surgeries, rehabilitation and adjustment that resulted from Bill's tragic accident had a serious impact on the lives of John and his mother. John managed to stay in school, coming to Toronto General Hospital to support his mother and visit his brother as often as he could. But these substantial absences affected his academic performance. Faculty administrators offered what support was possible to maintain John's progression through the program, but it was very difficult for him to remain a student while being a concerned brother and a supportive son. Just before John's final exams in April 1949, his brother returned home to continue his rehabilitation. John remembers this as a serious time that caused him to mature and realize the importance of moving forward on his dreams.

4
BUILDING A LIFE OF BUILDING

IN his role as president of the Civil Engineering Club for Ajax Campus in 1948-1949, John invited TTC General Manager and personal mentor Bill Tate to speak to the group at one of its monthly events. John's schoolmates were impressed, not only that Tate was there but also to see him exchange jokes with John in a lively discussion with other participants. The Toronto subway construction was underway at that point, in particular the central section of the Yonge Line between Dundas in the south and Summerhill in the north. The TTC was hiring widely. John notes, "Every kid in that room was interested in that, so he was a very popular speaker." Then, as now, many students rely on a decent paying job between academic years to enable them to remain in full-time study.

As his second-year studies were coming to an agonizing end, John was struggling with his brother's recovery, faltering grades and a waning interest in coursework. He knew that if he wanted work to look forward to at the TTC, he needed to make an appointment and walk respectfully into Tate's office and formally ask for assistance in his efforts to get a summer job on the track gang.

"What time would be convenient for you?" asked the eager student of the man whose grass he cut in the summer. Looking back on it now, he concludes that it was an audacious move. It amounted to asking to be treated specially when he was far from a remarkable student, and this was

Subway construction on Yonge Street in 1949.
(Photo: City of Toronto Archives)

Tate's workplace, not the backyard or a fishing trip. Neither youth nor fear held John back at that age, or later, for that matter. He planned his meeting well, realizing that his personal connections to the man might not distinguish him enough to penetrate the hiring regime of a union environment at the TTC. After a discussion about what John hoped to gain from working underground for the Commission, Tate gave the clichéd response, "I'll see what I can do."

John worked that summer as a track gang labourer on the section of the Yonge Street subway from Dundas station north to Summerhill. His pay was eighty-one cents per hour and rose to eighty-two cents by summer's end. "And I thought I had it made," he recalls.

John gained surveying skills through the work, one of the many practical components of construction that would be an entrée into jobs that followed, even prior to completion of his degree. The temptation to work and acquire skills that a university education would not provide was constant throughout his university years. It would win out, a year later, when excitement to work prevailed. In the meantime, with sufficient earnings to return to third year, John returned to school, albeit frustrated with academia.

In third year, a number of factors coalesced, prompting a typically single-minded and focused John to stray from his dedication to degree studies. He felt a rising sense of conflict, which prompted decisions that later would prove significant in his life and career path. Perhaps his shift in thinking was a response to Bill's near fatal accident a year before. For the first time, he faced the realization that no one's future is guaranteed and that life can be drastically altered by a single event over which one has no control. This made him even more committed to achieving his goals as quickly as possible.

John was still certain that he wanted to build and be part of a team on massive projects like the subway, ones that would change the landscape. He questioned the practicality of a diverse theoretical curriculum that bore little relationship to the world of heavy construction. The third-year course content reaffirmed his skepticism. His school colleagues likely also wondered what some of the courses they suffered through had to do with the jobs they would get one day. But it was amplified for John. "I was getting more and more convinced," he says, "that what I was being taught would never have any application to the real life of an engineer." This frustration, combined with the opportunity to participate in a technically advanced infrastructure project not far from Ontario, would eventually win out, undermining his commitment to complete his degree right away.

Although he was quite competent at labs and coursework, John could not understand why his exam performance was problematic. It was hard on his confidence. Among the relevant, timely and technically sophisticated courses that he completed were Cements and Concrete, Construction Surveying, Elementary Structural Engineering and Geology. But the relevance of Theory of Heat Engines was lost on him. *What could a civil engineer possibly need to know about heat engines?* he wondered. Perhaps today's civil engineering students feel the same way about studying the laws of thermodynamics. Even Modern World History and Political Science, also curriculum requirements, did not garner the disdain that Heat Engines did. The course was described, in the 1947 University of Toronto catalogue, as "the history and development of heat engines generally, the principles upon which they operate, and brief descriptions of the mechanical and thermal features of the different kinds of heat engines used in practice."

As often happens, a particularly difficult course stays with a student. In this case, the course has taken on nearly mythic proportions in John's life. To this day, Heat Engines is "the one that got away." As John saw it, "It was basically about combustion engines, and heat cycles, something I would only need in the maintenance sheds on the job site."

The Heat Engines course has become, together with other opportunities for ribbing, including his incomparable tenacity and receding hairline, the topic of many a replay throughout his life. His own father-in-law had a go at the joke when he secretly put toy engines in John's suitcase on his honeymoon. John takes it all in good fun, laughing more heartily at himself than anyone else does.

One of the consistent stories that came out of his university days involves John's passion (to this day) for peanut butter. His school chums used to joke that you could put dirt on bread and slather on peanut butter and Johnny would eat it. So, being engineers, they performed the experiment one evening in the common room. Some sandwiches were made, with John's including the contents of a recently swept room. As the onlookers smirked and tried to act casual, he ate his peanut butter sandwich with gusto. No one was sure if he caught on or not. It got a huge laugh from housemates and from John himself.

In that third academic year, John's studies and labs were on the main campus in Toronto. The change of location required new habits and induced some monotony, which, combined with increasingly frustrating course topics, challenged his commitment to complete his degree before going off to work somewhere. Although not an exceptional student in the academic sense, his ability to absorb information for practical application ensured that he drew the best possible education from U of T. The social and personal benefits of his university years left a more lasting impression. The friendships and kinship with fellow students, combined with more active participation with his fraternity, Kappa Alpha, were the glue John needed to complete his third year.

John's decision to become part of the Kappa Alpha (KA) fraternity, when the opportunity arose in second year, was purely social. Many of his closest lifelong friendships, including those with Bruce Mairs and John "Munk" Macdonald, were made either through membership in KA or in association with members, as was the case with Denis Evans. Certainly, at the time, it was thought that relationships from fraternity days might be a networking opportunity in working years.

But John owes far more than business deals to KA. The fraternity ended up being the element of university life that he feels personally most indebted to because, some time in third year, he went to a KA event at Cornell University and spent a few hours alone with one Margaret Campbell.

He knew Margaret reasonably well because she was the girlfriend of a fraternity brother, John Hardie. The two had often double-dated with him before the Cornell trip. Although John would bring a date for evenings out and dances, he rarely brought the same one twice. Sometimes just the three of them, the two Johns and Margaret, were together for a party or event. John just could not find a girl who had what he felt he needed in a partner.

Margaret, a student of the U of T Occupational Therapy program in 1950, had known John Hardie since grade 10. She was at St. Clement's

There was always time for laughter – and poker – when two of John's lifelong friends from university days Munk MacDonald (left) and Denis Evans (middle) came to visit on the James Bay project.

School then, and he had taken her sister out. In fact, when he and Margaret started to date, John Hardie's father used to jokingly ask him when he was going to take out Margaret's mother.

As John Bahen tells it, "We were all scheduled to go on a weekend down to Cornell University in New York State; Margaret was going, too, so was Hardie and so was I. While we were down there one evening, John asked me, 'Would you please take care of Margaret for a minute? I'll be right back.' I thought I could help him out by doing the job!" Ever the man of his word, John kept an eye on Margaret, and they have been together ever since – for sixty years of marriage at the time of writing. Their evening together ignited a relationship of remarkable strength, passion and permanence. It may have been one of the few occasions in John's life that took him by surprise.

Back in Toronto, in his typically direct manner, John went to Hardie to tell him that he planned to ask Margaret to marry him. Not that Hardie had a choice in the matter, but the courtesy was one he deserved.

Of Margaret, John says, "She was up on absolutely everything, and she knew about things. She was the smartest woman I'd ever met. I thought from the beginning that she was an intriguing woman." He also adds that she had the nicest car on campus (her father's).

His partnership with Margaret has been his most treasured. "Well, she has been everything that I wasn't, hasn't she? And some things that I was, she improved upon."

In April 1950, hoping to return to the TTC for a second summer, John again sought the support of Bill Tate, this time in pursuit of a position of some authority with the track gang: Survey Party Chief. Knowing that one job in the lead role was available each year, he requested, in a somewhat formal office visit, that he be considered for it. Tate reassured John that, if such a job were indeed available, he would have it.

"I wanted to get riding the rails," John explains. "I had a big machine that I rode along the tracks. It felt like I was meant to be doing that."

It was the appeal of the hands-on that had sustained him during the school years. The TTC work was pivotal in that it also gave him surveying skills, some supervision experience and interaction with project engineers who were implementing the latest construction methods to build Toronto's subway lines under existing streets and buildings.

The position as Party Chief was his, and John passed that summer improving his surveying skills and supervising three other students. While the work had previously offset his sense of frustration with school, it became more of a threat to school when opportunity came knocking before he returned to U of T for his final year.

An American engineering firm out of Chicago, De Leuw, Cather & Company, had a number of consulting engineers at work on the Toronto subway system and station designs. Through his role on the survey crew, John had regular contact with the company's engineers. They spoke about their other projects, one of which was the New Jersey Turnpike, of which their firm was the administrator. When Bill Tate mentioned to John that the turnpike job, which had a North American reputation as a megaproject at the time, was running behind due to a lack of skilled surveyors, it was not difficult for John to ask the American engineers how he would go about getting work down there. Although he may have been a well-advanced, soon-to-be fourth-year civil engineering student, his primary appeal was that he was a young hard-working man with surveying skills.

John sent a letter, with Tate's reference, to the De Leuw, Cather managers in Hightstown, New Jersey, requesting a job as a surveyor. That summer, he completed his work with the Toronto Transit Commission and hitchhiked through New York State to Hightstown with no intention of returning for fourth year right away. School would have to wait. It was time for twenty-three-year-old John to have some fun on a megaproject that was larger and more real than anything he had envisioned in the schoolyard sandbox.

5

SOUTH TO THE TURNPIKE

THE New Jersey Turnpike build had a reputation even before the first fleet of scrapers and graders took to the fields. The need for the project had been discussed at the state level for over forty years. While it was generally considered prudent for addressing congestion, while at the same time encouraging growth, its timing, design and layout were debated for years. On one thing there was consensus. The turnpike was essential to relieve traffic congestion on local roads between towns and cities that had experienced steady population growth since the turn of the century. It needed to be built, and sooner rather than later.

The turnpike had all the features of a job that, in later years, would draw John's attention when he was choosing desirable and suitable work for his team to bid on. It was a project built with high expectations of what it would do to improve movement, the economy of the northeast and the quality of life of people in both small and large communities. These expectations, added to the political sensitivity about the years of perceived delay, put the project in the spotlight. It was the project to watch and to learn from, because it was both a response to social and economic issues and an engineering challenge. It had a strict timeline, employed the latest construction materials and equipment innovation, incorporated environmental site challenges requiring creative and advanced engineering, was on a vast scale (for its time), was built by an innovative funding scheme and had a prominence in the heavy construction industry owing to its budget size. The turnpike was *the* job to be on if you wanted to learn the latest methods. John surmised he could learn on the job.

The 118-mile privately funded toll-road running from Delaware's Memorial Bridge to New York's George Washington Bridge was, and remains, a critical connecting link for interstate travel in the northeastern United States. Time to completion was a top priority for the New Jersey Turnpike Authority, created to oversee the expressway's development. To address the timeline challenge, Turnpike Authority officials broke the project into seven segments to be constructed simultaneously. Within each section, there were contracts and sub-contracts, each open to competitive bids. It was a complex system for project management, one that required precise, high-quality work and expert communication to achieve timing goals. The project completion period was some twenty-three months, from January 1950 through November 1951. The contractors worked under the motto, seen on signs throughout the project site offices: *"The turnpike must be done by November '51!"*

The project had the drama and cutting edge engineering challenges that John craved then and would continue to desire in every job that he chose to be involved in for the next forty years. The turnpike employed sophisticated design elements to eliminate the need for curves, left turns, stop signs, traffic signals and slow-speed sections due to grading severity. User safety was paramount in design, a reason why consideration was given to new surfacing materials. Those selected were heavier, skid-resistant and ultimately more cost-effective. The multi-lane turnpike was considered quite an engineering marvel for its time. It was characterized by innovative features such as wider lanes, a median, wide shoulders and reflectorized striping at the road's edges.

John arrived at a timely period during construction when, just as the De Leuw, Cather engineers in Toronto had said, slow surveying threatened to delay the roadworks. Surveying is labour intensive and demands precision by experienced teams. The survey crews on the turnpike were thinly stretched across all seven sections being built simultaneously. Perhaps an emphasis on construction and underestimation of pre-construction preparatory work are to be blamed. In any case, John was put to work on structure surveying immediately. Although there were some massive bridge spans on the turnpike job, the demanding smaller structures, some two hundred forty of them, including small bridges, cattle diverters, culverts, ramps, tunnels and abutments, were the surveying afterthoughts that kept John employed. What he felt might be a few months of work extended easily into the spring of 1951.

The work was much more challenging than John expected. He worked three weeks without a day off, double shifting every day (or so he says). The interchange work was weeks behind, whereas the roadwork was running nicely. The pressure to accurately and quickly survey each stake was immense, because forming and concreting crews were coming along immediately after to do their part. Survey work had to be exact and correct the first time. This same "just-in-time" tension existed across every contract and every segment of the entire turnpike job throughout its construction period. Other people – crews and contractors – relied on the timely and precise completion of each task.

John ensured that each survey stake was in the ground and correctly marked, accurately describing what point it was and what structural element it represented. All dimensions, headings and right-of-way limits needed to be precise and recorded meticulously. There were markings for foundation points, footings and corner points. Almost immediately after each footprint was staked, an excavating crew came into the trench around the stake and began digging from point to point. It worked like an assembly line process. Although John would achieve the staking of an entire pier of a bridge relatively quickly, before he was entirely done, the next stage of construction would be beginning. He was learning rapidly, and being part of the team was a powerful rush for him, but there was no time to rest and appreciate the massive undertaking. Nor was there a moment to realize that his life was taking a firm direction toward construction.

John lived in Hightstown, not far from Princeton, New Jersey, and quite close to the turnpike. He roomed with John De Vany, who worked with him on his first survey crew and later became a site superintendent. De Vany would be a groomsman at John and Margaret's wedding less than two years later. John recalls reporting to a competent young engineer, Robert Richards, when surveying ramps. Richards would rise in the company to become president of De Leuw, Cather & Company until his retirement in 1979. Richards was one of the principals in the design of the New Jersey Turnpike and an early inspiration to John.

When a fellow he roomed with mentioned that one of the contractors needed someone to run an asphalt plant, John's interest was easily diverted, especially after working such long hours for weeks. Hearing that he could earn three times what surveying paid, he took on the job of running a large asphalt plant – actually four units running in tandem – for George Brewster and Sons, one of the major contractors on the turnpike. As it happened, another major contractor on the turnpike was Peter Kiewit Sons Inc. of Omaha, Nebraska. John had heard the name before, as the company was also a joint venture contractor on the Toronto subway. He made mental note of it. There would be many such reminders of Kiewit's presence and status in the early years of his career.

One day, when John was in the plant, the United States Army Corps of Engineers came to observe the use of the latest methods and materials for roadworks, hoping to apply what they learned to the design of a series of air base runway upgrades planned as part of the Cold War preparations. Following that, they asked John to meet with them in New York City, at which time they offered him a job in Algeria, North Africa.

The Army Corps was engaged in air base upgrades in various locations, including northeastern Canada and Africa. Although the location formally offered would not have been John's first choice for a position with responsibility for runway materials, he was prepared to go. He decided that working on military sites for the Americans would be an experience that likely would not come his way again. He was already living away from home, and he had missed a good part of the fall semester, which added logic to his rationale for accepting the move from New Jersey.

Somewhere between his final days in Hightstown and his arrival at an American Air Force Base to board a plane for Algeria, John went home to Toronto. He had a plan.

He arrived at the Campbell home with a ring in hand and proposed to Margaret. They had been apart almost a year, but each was stalwart in their commitment to be married to each other. It was an offer that, while not unexpected, was odd in its timing because Margaret would not see John for another long period. Nonetheless, she accepted, and within hours, John was off to his new job.

Kay Bahen thought Margaret was the perfect choice for a daughter-in-law, but she worried that John might be overly concerned with construction and be less than the attentive husband Margaret deserved. This had been a reality of the early days of their courtship and continued through their marriage. As Margaret acknowledges now, she always came after construction and sometimes, with her approval, after the children. It was also a reflection of the generation and the time. Progressive and independent, Margaret had been working full-time in occupational therapy on the quadriplegic ward at Toronto's Sunnybrook Hospital when she and John met. Once they were married, she continued as a volunteer for another couple of years. It was understood that her ultimate role would be to support her ambitious husband and, eventually, raise their children in his predictable absence. She knew well what she was committing to. She loved John and passionately assumed a role as his partner as well as wife and future mother of their children.

In September 1951, John reported to the point where staff and civilians were boarding Military Air Transports (MATs). However, when

*John and Margaret at her home on
Strathallen Boulevard the night they became engaged.*

Army clearance and security staff discovered that John Edward Bahen was a Canadian citizen with only a valid Canadian passport, the prospect of the exciting job and challenging responsibility was looking far less likely, leading instead to the possibility of his name being added to the American draft list.

The issue was quite complex. First, only American citizens could be transported overseas on MATs. Second, the United States was at war with Korea. John had heard that recruiters visited the turnpike construction sites pursuing the rumour that men avoiding the draft were hiding out on the construction teams. They acquired John's name as one of many eligible to serve, assuming that he was American. The fact that he was Canadian did not prevent his inclusion in the draft either. He had been residing and working in the U.S. for a sufficient period to be labelled a resident alien. At a time when every service-eligible citizen was being accounted for by Army recruiters, there needed to be a substantial reason why they should not add John's name to the draft list and require him to do minimum training and service.

However, the Army Corps' desire to have John on their airport construction site was more powerful than the insistence by the other arm of the Army that he be drafted for service in Korea. So his health status was changed to 4F to allow him to be transported, such relocation being limited to within North America.

Algeria was replaced by Goose Bay.

6

NORTH TO GOOSE

JOHN was twenty-four years old when he landed, by United States Army jet, in Montreal, Quebec, one of the stops on his way to work at the isolated military installation at Goose Bay, Labrador. He had worked in New Jersey for a full year and was headed to "Goose" for work that ultimately would be the reason that he postponed his degree completion for two full academic years. School was both literally and figuratively far away, but his path toward the work of his dreams was proceeding apace.

John knew he would benefit in the present and future if he gained experience that could be provided only by hands-on work on a massive construction project. He would attain extensive knowledge of materials, types of fill (concrete and asphalt mixtures in particular), stability and compaction testing, and excavation and earthmoving – expertise that would inform his work for the rest of his career. He felt that this was not something coursework could offer, and, on the basis of this argument, he obtained approved leave from his studies. With the promise of the Army Corps to train him and provide him a testing lab and some staff, he took on the responsibility of ensuring that the earth being moved and placed under the Goose Bay runways could be compacted adequately and that it would be structurally sound for years under the use of the largest military aircraft carrying massive payloads.

The Goose Bay base was built beginning in 1941 by the Canadian government, but it was also home to both the United States and Great Britain during World War II, through lend-lease agreements. By 1950, the base became part of the Northeast Air Command, at which time it was used as a refuelling and last stop for planes heading to Europe. The United States was engaged in a "cold war," a moniker that described the tense political relationship between the Union of Soviet Socialist Republics (USSR) and the West. In the years after World War II, efforts to protect against the USSR – an enemy closer to home over the Arctic Circle – became the strategic focus. Eastern Europe did not pose an independent threat any longer, but the USSR, of which it was a part, in its self-appointed role as leader of the weakened the Eastern European countries, posed a formidable challenge, especially given the evidence that development of nuclear arms in Russia was on pace with that in the U.S. Operation Bluejay, the establishment of military bases in Canada's extreme north, as well as in the Arctic and Greenland, was key to maintaining a state of readiness. Goose Bay was one of the strategic forward bases, considered a staging point in co-ordination with the base at Thule, Greenland, that fulfilled a similar role. In 1957, Goose Bay became part of the Strategic Air Command for the United States Air Force. Because of the then top-secret existence of atomic bombs at some of the bases, John had to sign an agreement to never discuss this with anyone – which he didn't, not even with Margaret.

John reported directly to the commanding officer of the United States Army Corps at Goose Bay, Colonel Willard Roper. Thinking back over the jobs that he has undertaken directly or as a project manager, John acknowledges that, given his lack of experience, the excessively high expectations, the powerful client he worked for and the critical impact of making a mistake, Goose Bay may have been his most difficult. Having so much responsibility at such a young age weighed heavily on him.

Little of John Bahen's work life has not been challenging. Learning while doing is part of his drive. John would learn that nothing ever quite comes easily, nor does it ever go as smoothly as it might. However, his ability to draw the best knowledge and experience from every event is at the core of his success. Each successive role was logically possible because of something that he had learned or proven already. He had gotten his hands dirty doing everything from excavating and surveying to running heavy machinery. He had three years of formal engineering schooling and experience constructing the subway and streets in the Toronto area. He never doubted that he could do whatever was asked, but he certainly wondered why such faith was placed by the Corps in a kid from Canada about whom they knew very little.

Asked what skills the Army Corps believed he possessed that made him worthy of such responsibility at his age, John jokes characteristically, "Maybe I was the only guy that didn't complain about the food!"

Then he explains more candidly, "All I wanted to do was work, work, work. I was able to control their operations by putting a stop to activity if we weren't getting the quality we required for construction."

The base at Goose Bay and the town of Happy Valley had a combined population of about five thousand in 1951. The Canadian Department of Mines and Resources had surveyed the area for the Royal Canadian Air Force, ensuring that a series of preferred geographic features were present for the base. The base lands sat solidly on a sand plateau above the flood plain where the Churchill River empties into Lake Melville and far enough inland to avoid the climatic influences of the Atlantic Ocean. The harbour in Hamilton Inlet offered good anchorage, with navigable waters from Terrington Basin into the lake. It was an engineering challenge to build on a plateau that had sand deposits over three hundred feet deep. This required that constructed elements be on a densely compacted subgrade to ensure minimal impact of settling on the settlement-sensitive infrastructure to be built: runways, taxiways, hangers and utility connections.

Some of the old, shorter asphalt runways had been demolished in order for newer ones to be constructed. With the advent of B-52s and B-26s, heavy payloads were the norm, especially at a military base acting as the last refuelling stop for flights overseas. John's responsibility would be to manage the material testing lab and its small staff, and the analysis of the various material samples taken and cured (in the case of concrete cylinders) on site. The Corps of Engineers sent an engineer from Mississippi to stay with John while the materials testing lab was being constructed. That is the only real training he received. Six months after his arrival in Goose Bay as "just an engineer," he was promoted to Materials Engineer and assumed his role at the Materials Testing Lab.

John was a constant presence near the points of construction while samples were drawn for analysis. On-site testing consisted of fitting a cone into a tubular hole in the sand and drawing a sample of the compacted sand, which was weighed against an identical sample of uncompacted sand to determine the degree of (extent of) compaction. This was a continuous process of evaluation. Construction took many weeks. The main runway at Goose is over eleven thousand feet long (still one of the world's largest), and the secondary runway is over nine thousand feet. At the time of the upgrades that John was involved in, Goose Bay was called the busiest airport in the western hemisphere. This was mainly due to its location as a last stop before flights headed over the Atlantic, as well as the fact that it was used simultaneously and co-operatively by many air forces.

Other challenges included the production of quality aggregates for base course materials, concrete and asphalt products. The local granular materials required significant processing (crushing, screening, washing and blending) to meet design specifications. As they were of particular importance to the project's quality objectives and schedule (there was no time to rework deficient outcomes), John took it upon himself to check the gradation or sizing of the aggregate arriving from the processing facilities. He made batching (mixing) adjustments on the spot as required. Samples taken from the site casting were cured for twenty-eight days and destructively tested. A critical element of John's responsibilities was ensuring that his personnel followed proper sampling and testing procedures, that laboratory equipment was calibrated and that the lab not only was well maintained but also looked well maintained.

John was elated to be given the opportunity and responsibility of Materials Engineer, but he was always wary that he might not be seen by others as qualified and wondered fairly regularly if the administration might suddenly decide that they had made a mistake in giving such responsibility to a young man who hadn't even completed his degree.

During construction at Goose Bay, John again heard the Kiewit name, in this case in connection with work being done at the base at Thule, Greenland.

"While at Goose I had a lot of time alone to think about my future. I was looking to the future always being in the construction industry – but where in the industry was the question. The main work going on across North America then was airport construction, after that it became roads and after that energy and hydroelectric power. My choices were getting limited. It looked like I was going to be in roads or in the hydroelectric business."

Goose Bay was both a head start and long-term job security for John. While acquiring skills and experience, he was able to demonstrate he could shoulder responsibility. Perhaps as important, he built a good reputation as well as industry connections that put him in an advantageous position when applying for work against other engineering graduates who lacked such experience. These factors would qualify him for leadership roles earlier than others with just academic qualifications.

~

It was a rush for John to get back from Goose Bay to Toronto in time for pre-wedding festivities in August 1953. He managed to play a final hockey game on the outdoor rink at Goose a couple of nights before his flight. He was a key member of the Army Corps team, and this loyalty kept him on the base perhaps later than he should have stayed. It took two military flights and a commercial flight to get him to Toronto, after stops in northern Ontario and Winnipeg.

John had been away a total of two and a half years. By the time he came home for the wedding, he had visited no more than three times and only for a few days each – one trip being for his brother's wedding the year before. This ability to withstand isolation and endure separation bode well for John and Margaret's future relationship. It would be forty years before the travelling for work slowed down again.

Margaret's father, Alex Campbell, a decorated World War I veteran and founder of Acme Carbon and Ribbon Company, would eventually become both mentor and father to John. Margaret recalls that, in the beginning, her father called John "the guy in the ice cream suit," because of a blue and white seersucker suit he wore. She took that as a sign that her father did not take John seriously. In the early years of their relationship, Mr. Campbell voiced concerns that John might never finish his degree. He understood well that the thrill of challenging work and good pay might sidetrack John from his last year, and he was of the strong opinion that the degree should be completed. However, while Mr. Campbell was not so sure that they would work out as a couple, he had no doubt about John's drive, nor about Margaret's support of him. Nor could he have foreseen the incredible partnership the two of them would forge. Over time, he became the most stalwart supporter of John's career.

Margaret's mother, Edith, whom John affectionately called "Gert," was fond of him from the beginning and responded warmly to his lighthearted personality and love for her daughter. And Alex Campbell had a high opinion of John's mother. He thought her a brilliant businesswoman, admiring her ability to run the Eaton's restaurant and then the exclusive Childs.

The wedding on that hot August 29, 1953, was featured in great detail in the social events column of the *Globe and Mail*.

John and Margaret on their wedding day, August 29, 1953.

Partial write-up of John and Margaret's wedding in the Globe and Mail.

Mr. and Mrs. James Alexander Campbell
request the honour of your presence
at the marriage of their daughter
Margaret Edith
to
Mr. John Edward Bahen
on Saturday, August the twenty-ninth
nineteen hundred and fifty-three
at half-past seven o'clock
St. George's United Church
Toronto

and afterwards
The Granite Club

bell, to Mr. John Edward Bahen, son of Mrs. Katherine Bahen and the late G. P. Bahen.

The bride's handsome gown was an original model of champagne French Chantilly lace over many layers of tulle, over taffeta. The portrait neckline was heavily encrusted with opalescent sequins and pearls, and the sleeves were long and pointed. The bouffant skirt had godets of accordion-pleated nylon tulle. The bride's veil was caught with a tiny jewelled cap of matching lace, and she carried a semi-crescent of gardenias and stephanotis in champagne tones.

Given in marriage by her father, she was attended by her sister, Mrs. Hugh E. Zimmerman, matron of honor, Mrs. Hubert A. Gray and Mrs. Thomas R. Clarke. They were gowned alike in copper parchment shantung over taffeta, fashioned with wide square necklines, softly moulded decolletes and full skirts. All carried baskets covered with the shantung, filled with copper-toned chrysanthemums and cleosia, and wore semi-circular headdresses in the same floral effect. Mr. William G. Bahen was groomsman for his brother, and the ushers were Messrs. John De Vany of Chicago, M. Bruce Mairs, Robert Bertram, and A. M. Campbell, brother of the bride.

After the reception at the Granite Club, the bride and groom left for "Pink Sands," Harbour Island in the Bahamas. They will live on Old Forest Hill Rd.

It was held at St. George's Church on Lytton Boulevard. Margaret's matron of honour was her sister, Mora Zimmerman, and her bridesmaids were Pamela Gray and Barbara Clarke. John's best man was his brother, Bill, and the ushers were his friends John De Vany, Bruce Mairs, Bob Bertram and Margaret's brother, Alexander ("Mac"). At the reception for three hundred guests at the Granite Club, John spoke extremely well, impressing both Margaret and her father.

Colonel Willard Roper (John's commanding officer at Goose Bay) was instrumental in ensuring access for the couple to a luxurious honeymoon at Harbour Island's Pink Sands Resort, at the time a members-only retreat in the Bahamas. It was a short but romantic launch into their lives together after years of letter-writing and very little time together. It was also an escape from the university and construction life that they each had been absorbed in. Even on his honeymoon, John was reminded of what awaited him at home. His father-in-law's way of hinting that his engineering degree needed to be completed upon his return was to plant some toy engines among lots of confetti in John's suitcase.

7

LEARNING ON THE JOB

THERE was no starting back to U of T for John unless he conquered his demon, which took the form of the Heat Engines course exam. Although somewhat supportive of his departure and hiatus from academia while working, the administration of the faculty insisted that he pass the Heat Engines supplemental exam to restore his good standing, which he did. He was told by a professor that it was the exact exam that he had failed years earlier. "Funny," John replied to him, "it didn't seem a damned bit easier!" Re-enrolled in the fourth-year courses, now with experience, a wife and a very driven attitude regarding what he would do after he officially became an engineer, John completed the year with relative ease. Ironically, the day he graduated, the Heat Engines course was removed from the core Civil Engineering curriculum.

~

McNamara Construction Company had done work at Goose Bay in the 1940s. When its president, George McNamara, was revisiting the site to possibly tender on some work, during John's time there in 1952, he was introduced to "that young Canadian guy" who was working on the runways. With a handshake, he told John that, after he completed his fourth year of school, there would be a job for him with McNamara. John held him to it.

As an early contractor at Goose Bay, McNamara realized that John's experience working for the Army Corps spoke highly of his ability to achieve results to exacting standards in the context of harsh site conditions. John was thrilled to be offered his first job at home in Toronto. Married less than a year, fresh from completing his engineering degree, he was poised to launch into his building career. "It was my first 'real' job," he says.

McNamara was a small but growing family-run operation with a competitive attitude when John started with the company. The founder, George McNamara, had three sons, George Jr., Bill and Paul, who ran the company together with their cousin Harold. John was with McNamara for over ten years, work that saw him gain local and international experience in two main types of work: roads and airports. But in the most challenging of the projects that he managed at McNamara, he acquired his first hands-on challenge with a hydroelectric power structure, something that he would come to base his career success upon. The McNamara years can be credited with providing John with his intricate knowledge of heavy equipment, his site-based, involved approach to project management and something he simply calls "people skills."

In his early work with McNamara, he accepted increased responsibility eagerly. He was a superintendent, managing provincial Ministry of Transportation roadworks, including some on sections of Highway 401 (the McDonald-Cartier Freeway) through the City of Toronto. This kept him at home most days, a luxury his career would not afford for long.

John had been involved in the massive earthmoving job at the Avenue Road Bridge where he first met John Wilkes, who would become a friend for more than fifty years. Wilkes, the provincial planning engineer responsible for municipal highway and bridge contracts, is credited with the idea of having combined multi-lane collector-distributor functions on major highways, resulting in the twelve-lane span of express and collector lanes on today's Highway 401.

Of their meeting, Wilkes recalls: "John started at the bottom. You need to know all the jobs, and John made sure he did."

Wilkes is one of the few men who truly have known John Bahen since the beginning of his career. While they both studied Civil Engineering at U of T, Wilkes graduated a few years before John. They share common sentimental feelings for the place. The two regularly encountered each other on jobs in the Toronto area, and, together with their wives, they became lifelong friends.

"There were no problems with his ability. There was nothing we didn't like about him," says Wilkes.

John was noticed even then by many for his rapid rise in the McNamara company.

He was something of a maverick because he was an engineer who chose to be hands-on in the field and observe the progress of the construction work. The design fascinated him, but how to build it was his true passion. Today it may be more commonplace – and even the expected norm – that an engineer needs to be competent from the technical to the construction site, but it must be recalled that John came through engineering during the first years, post-war, when the profession was transitioning from the administrative to the technical. The "business" of engineering was still a desk job to most, a shortcoming that John has been critical of his entire career and one that he was never willing to settle for. As he puts it, "You can't plan from an office and expect something to work in the dirt."

Most of the superintendents were very experienced construction people, not engineers, so John stood out among them, offering an employer the best of both worlds. "He was young, eager and pretty damned good, on top of everything," Wilkes recalls. He learned an immense amount about the intricacies that skilled staff and timing of equipment have on a job – knowledge that he applied to every bid he worked on for the rest of his career. Ontario highways were fertile ground for a budding builder. Wilkes knew hundreds of superintendents, project managers and engineers in his long career with the Province of Ontario. Reflecting on what John ultimately accomplished in his career, Wilkes says: "He had to be something – you have to be quite unusual to start off at the very bottom and now to be someone that the world is talking about."

John, asked why McNamara had such confidence in him in those early working years, puts it down to his willingness to learn and ask for help. Wilkes would call this a huge understatement.

Among the roadwork projects, where John got a taste of how the people-machinery-materials triad will constantly challenge a project engineer, he recalls a particular section of Highway 401. When McNamara was driving a sewer tunnel from St. Clair Avenue and Spadina south to Bloor and Spadina, no major surprises were expected. John thought that they had likely seen one example of every conceivable issue by then. This was the legacy of that job for John, to expect the unexpected, because you cannot afford to be complacent on a job site. With its experience in earthmoving, concreting and general roadworks on the 401 across Toronto, McNamara Construction was confident in knowing the men, machinery and materials in the area. But they could not know everything. When an underground gas explosion occurred during construction, it was John's responsibility to deal with it. The explosion was extensive enough to have reportedly cracked the foundations and walls of nearby homes, in addition to causing a sinkhole on the excavation site. John took it on as a learning experience and walked every foot of the damage, which included going into some of the area homes where claims against McNamara or an insurance company were going to be made. To his shock, the presence of methane gas was refused as an insurance claim on the basis of some potential compromise of a nearby sewer tunnel, and no reimbursement for lost time or additional work was made.

As the project was completed, he and Margaret were only a couple of months into enjoying their first house, in Thornhill. John arrived home and announced matter-of-factly, "I have to go to Carleton Place. You stay here and sell the house." On top of that, Margaret had just had the first

LEARNING ON THE JOB

of what would be many miscarriages. She sold the house and moved to Carleton Place, just outside Ottawa, where they lived in a trailer, courtesy of McNamara, on a gas station lot. The trailer was so cramped that their large collie, Shane, had to run to the end of it and jump up on the bed in order to turn around.

John's promotion to managing the roadworks section of the Trans-Canada Highway through the town of Carleton Place came with increased responsibilities. The first contract amounted to about fifteen miles of highway work. This was the beginning of three years in eastern Ontario. He remembers this early work fondly for its variety. There was excavating, paving, bridgework and a lot of traffic management. This was the work he had been waiting to get at, and he was superintendent. When the Trans-Canada highway work was nearly complete, the St. Lawrence Seaway contracts started coming in – McNamara was successful in obtaining some nine in total – and John became area manager. The contracts included six large road projects, the Cornwall Bridge and still another section of Highway 401 to the provincial border with Quebec.

John sums it up like this: "I learned many things. First and foremost, I had to learn how to direct people and keep them busy and keep them well directed. I had to detail plan."

The scale of the St. Lawrence Seaway in eastern Ontario was unprecedented. Its purpose was twofold: to build three dams as well as two powerhouses to harness the river drop and to replace the existing (and dated) Cornwall Canal in order to allow the growing number as well as increased size of ships to circumvent the Long Sault Rapids.

During this project, the Bahens lived in a cottage on the Long Sault Parkway near the town of Moulinette, which, like several others in the area, had to be moved to avoid being flooded by the seaway. Houses, churches and churchyards were all moved to a new town farther inland. John was very busy during this time. Every hour counted, so it was not unusual for him to bring up to twenty-five men on the job for lunch so they could continue planning while they ate. "The butcher in Moulinette loved me!" Margaret says.

When the Long Sault Parkway was finished to Cornwall, they moved to the top floor of a motel on the seaway route in Cornwall where they could look the crews of the boats in the eye as they passed. "Crazy, but fun," adds Margaret.

The St. Lawrence Seaway project gave John the opportunity to show his talent for finding significant additional profits on a job, in this case by finding a cost-effective way to excavate the glacial till materials.

Each challenge on the job, whether with staff, equipment or materials, was part of a cumulative knowledge base that developed John into an incredibly resourceful project manager. He never made the same weak assumption twice, because he had the experience of what had happened on an earlier job to learn from. Gifted with a meticulous memory, once he learned something the hard way on the job, be that through lost time or financial impact, it would never happen again on a John Bahen job.

"I was learning as I went," he admits. He was also climbing rapidly as he went. The facts of his swift trajectory speak for themselves, despite his humility. From the time that he was hired by McNamara in 1954, he followed this progression in about six years: Project Superintendent, Area Manager, Grading Manager, District Manager and Vice President.

Neither school nor his work experience between 1950 and 1953 had given him skills on this scale; he was learning not only to manage but also to plan and, in some cases, how to correct errors of judgment with both people and plans and still maintain quality work and efficiency on schedule.

Another particularly memorable job was concrete deck work for the subway track suspended under the Prince Edward viaduct (also called the Bloor Street viaduct) main arch-truss bridge section on which he was project manager. John had grown up admiring the scale of the Prince Edward Bridge, the massive structure that is more than one hundred thirty feet above the valley floor. It was an engineering marvel, constructed between 1915 and 1918, based on a forward-thinking design that anticipated subway trains (or streetcars) eventually running underneath the vehicular portion of the bridge. The design included a lower deck of ironwork with structural strength and ample clearance for its use some forty years later for an underground train link from one side of the Don River Valley, at the Rosedale Ravine, to the other.

Working above the valley and over the river and roadways, John co-ordinated work that would enable the newly extended eastern subway line to connect the eastern and western portions of the city. It was a meaningful project in the life of Torontonians as well as for John, who had worked the summers of 1949 and 1950 on the first line, the Yonge Street subway line. The TTC's commitment to the subway was also the legacy of his mentor, Bill Tate, whose support and encouragement John had counted on since his high school days. John had heard Tate speak many times in private about his personal dream to see transit connect the two parts of the city.

A reinforced concrete layer more than thirteen hundred feet long needed to be poured on previously constructed steel beams, with more steel being added for reinforcement to create a track bed. John knew the schedule for the project was very aggressive because special steel had been ordered by the TTC. In the interim, he worked on innovative ways to process other parts of the job so he could make up for the lost time that might occur. He devised a way for concrete to actually be precast in the new subway and shunted along the existing base track to the area where workers, suspended on safety lines, could guide its payload into place.

"We never lifted it at all," he explains. "We just had little mine cars. We shunted them into place down at the level of the work and put the precast concrete onto them and away they went across the valley."

This was also the time of his first on-the-job encounter with Kiewit, which was constructing portions of the subway west of Yonge Street while McNamara was building on portions east of Yonge. For John, it was an exciting and challenging period of momentum with the steep learning curve of bids, contracts, jobs and opportunities.

―

The late 1950s were demanding for John and Margaret. By the fall of 1957, after four years as newlyweds, they had lived in an assortment of places around Carleton Place and Cornwall and had an apartment in Toronto, just north of Eglinton Avenue, for their belongings. In the fall of 1957, they purchased a house at 118 Yorkminster in North York in order to have enough room for the family they planned to have. A son, Stuart, had arrived in April that year, followed, in December 1958, by Susan. In those years, work sometimes took John away through the week, depending on the job he was working on; but he was, ideally, home most weekends.

In the meantime, John's mother, Kay, had been working as a house mother at a University of Toronto sorority. Prior to that, in failing health for several years, she had stayed with John and Margaret for just under a year when they were in Thornhill as she recuperated from heart ailments and cancer. She died in September of 1959, which devastated the Bahens. Just over a year later, John and Margaret's youngest, Michael, arrived.

8

BIDDING FOR AMBITION'S SAKE

JOHN did not take management seriously when they hinted that he might be needed to lead a team to Pakistan and conduct research for a bid on the Mangla mega-dam. To him, this was unrealistic for a company the size of McNamara Construction. Not only was the project anticipated to be one of the largest earth-filled dams in the world (it still ranks in the top ten), but it also was in a part of the world where McNamara had no experience. German and British firms were bidding in addition to the Americans, and McNamara was asked to do so because Commonwealth representation was required. But the excitement was intoxicating for John and the others. They had not yet faced the harsh realities of months abroad. The decision to be the lone Canadian company bidding on part of Mangla was an early hint to John that, at McNamara, ambition outranked capability.

While John and a team of six prepared to go overseas in 1959, the McNamara family managers also committed to expansion into the U.S. market, seeking to capitalize on the massive amount of construction work there, particularly in California, where a decade of major public works funding was beginning. John's involvement in the experimental expansion into that state was minimal for the first couple of years but would become quite intense later.

John did not think that the desire to reach farther afield for work, past local, provincial and national boundaries, was a smart business move for McNamara. To him, it indicated that the company executives were out of touch with how construction really worked. "To build in a place, you have to know the soil, the climate and what experiences previous builders have had," he says. "You need to know who to hire, suppliers and transportation costs, and you can't assess all that well enough in advance of a bid to get the costs and the price right. It's a big risk spreading out too far."

Among the estimators travelling to Pakistan was Ulysses "Shake" Hodgins, a man whose skills John trusted and whose knowledge he counted on. Preparing estimates with other construction sub-contractors who would contribute to the larger effort kept the men away for months. Three of those months they camped in tents, with armed guards posted by the tent flap. A great number of companies had sent engineers and specialists during the pre-bid period, and they were housed in a temporary camp where gunfire was heard, due to some local unrest. The Canadians were sternly advised to have some protection – hence the guards.

The rest of the time away involved information gathering and site investigation and research before bid preparation could take place with a larger joint venture partner, Taylor Woodrow in England. John explains the process in the following way: "Bear in the mind the difference between bidding and estimating. Preparing the bid, or bidding, is really only one quarter of the work. The other three quarters is estimating, planning details and putting costs against those details. And if you don't get the estimate right, it doesn't matter how well you execute it later."

Margaret recalls that the year John worked on preparing the Mangla bid he was home in Toronto only eleven days. And on one of those visits, Stuart was so excited to see him that he ran right through the cottage screen door. Although it was just a two-day visit, John managed to pick up mumps from the kids.

While in Pakistan, John spent a great deal of time touring other construction work to get an idea of available labour and accepted productivity rates. He was shocked by how poorly labourers were paid – a fraction of what a North American earned for similar work. In addition, in Pakistan, they worked in relentless 120°F heat. John saw first-hand the direct relationship between low productivity and incommensurate pay. He came to understand that motivating people required three components: a balance of pay, a goal that requires some effort and ongoing support from management. The absence, in Pakistan, of all three components greatly influenced his developing leadership style. This was a key piece of learning for him, one that he could never have learned in school.

In 1959, options for return trips to Toronto from Islamabad were few and randomly indirect, be it via Rome or Hong Kong. Knowing that, upon his return to Canada, he was needed immediately in Churchill, Manitoba, for a big airport site job and would be going directly there, he purchased some fresh clothes in Hong Kong, including a custom-made wool suit for $35. Pleased with its low price and speedy delivery (less than two days), he wore it to board his flight to Canada. The plane landed in a downpour and, as he ran across the tarmac, he did not immediately notice that his new wool suit was shrinking rapidly. By the time he reached the job site, the jacket sleeves and pants had shortened by about four inches. He laughs about this story still, because it reflects his lack of focus on personal details such as his clothing, compared with his absolute passion for his job. To add to his poor-choice-in-clothing story, he had purchased a lovely outfit for Margaret in Lahore: an Indian-style long top and wide pants. It was so big she never wore it – indeed, he laughed himself silly when she first tried it on.

For his project in Churchill, he was the first man on the site. The Cree Indians who had been hired to clear brush were friendly enough and introduced themselves. Chief Campbell was in charge, followed later by Chief McDonald. Their names struck John as rather unusual for Crees. History played a part here, as they were the descendants of earlier Scottish settlers and Aboriginal mothers. He immediately phoned his father-in-law, urgently stating: "Send membership sign-up documents for the Campbell clan. I have found the last tribe."

Meanwhile, the Mangla bid was not successful. That was not a problem for John. In those days, the firm only won about one out of ten jobs, leading him to reflect that he "learned to be a loser." He may have learned in the academic sense, but he never learned to like it. No one who worked with John over the decades could deny that he couldn't stand to lose at anything.

California was inviting in the 1960s, when a large number of contracts in the Feather River complex were being tendered. With George McNamara chasing ambition more than work, John was sent back and forth to California for just over a year to deal with thirteen troubled contracts.

The project manager that McNamara had in place had little American experience and was not successfully balancing the needs of the roadworks contracts with the largest project, a hydroelectric power job. In John's estimation, this man was more of an engineer – a pencil-pusher – than a builder: a shortcoming that showed in his management of the challenging contracts that required a construction-minded skill set. Although John was not yet forty, his experience frequently outweighed that of men twenty years his senior, as was the case in the California projects. Several contracts were losing money when he went to California in 1964. The Dam Underground Powerhouse (now named the Edward Hyatt Power plant) job at Oroville was a first for him on many fronts: his first profitable job in the United States, his first hydroelectric power structure – and his first time being shot at.

That story goes like this. After John replaced the project manager, he had to assess and address all of the challenges of each California job with the intention of raising them into a profitable situation. This was a monumental task for the powerhouse job because, in addition to staffing, machinery and materials challenges, there was anger building in the union. This led to personal threats against John. The union attempted to intimidate him by surrounding their leader with men when discussions took place. These discussions led to verbal threats on more than one occasion. But John was not one to back down on policy, approach or the plan.

Thefts of small equipment and supplies from the job site had also been reported. The man who ran the warehouse was able to monitor thefts by union workers. John's reaction to the theft was to confront the head of the local Carpenters' union. He had a good idea who was leading the thieves.

"The American workers thought they could pull one over on the laid-back Canadians," he says.

The morning after the confrontation, as he was having breakfast in the diner he frequented daily, one where he met with his superintendents and staff to review each day's work, his car, parked out front, was shot at. The California State police were alerted, because John had earlier reported the aggravated talks. John's reaction to the shooting was classic, unflappable Bahen.

"They just wanted to hit the car," he says. "If they wanted to hit me, they knew where to find me. It was just a scare tactic. I wasn't scared. I knew who I was dealing with. It was an education for me for later, though."

Police were posted outside his motel at night for a few days until the company (or the police) could resolve the situation. John ran across the union head in the parking lot and mentioned that his car had been hit by a stray shot. John laughs as he recalls the union guy's reaction, which was as blasé as his own: "Maybe you shouldn't eat breakfast there anymore." "Good idea," he responded.

And that was the end of both the thefts and the union problems. The lesson he learned there, which he would apply innumerable times in the forthcoming years of his career, was that sometimes the best reaction is no reaction at all. Not long after, McNamara was able to prove a connection between the thefts and the individual responsible. When the man was fired, frisked and walked to the gate, there was a lot of anger on site, but never any more violence.

In addition to the other firsts, Oroville Powerhouse was also John's first time in court, but unfortunately not his last. The Oroville Dam was a huge, magnificent structure. Although he admits now that a reasonable lack of experience plagued the team at multiple levels – after all, that was his reason for being there – they found and solved a massive materials issue that resulted in a court battle for cost recovery.

Drilling and blasting were taking place in the large cavern under the dam when blasting experts – "dynamite guys" – realized that the geological composition of the rock was such that they could not control the extent of blasting. When more rock than anticipated easily fractured and broke out, McNamara Construction had to replace it, at their cost, with concrete.

John, on behalf of McNamara, claimed that the rock overbreak was a change of conditions and submitted a request to the owner for compensation for the additional concrete. The client, the State of California Department of Water Resources, represented by Alfred Golze, the Chief Engineer, vehemently disputed the request. McNamara argued that the rock's tendency to overbreak should have been established as a geological fact in pre-bid materials. John had checked and rechecked the material details in the pre-bid package numerous times. John credits one of his colleagues, Sandy MacDonald, with assembling evidence supporting McNamara's claim.

A law firm in San Francisco argued for McNamara. The two sides could not agree on the amount. John was relentless. He spoke with Golze personally each day, updating him on the site situation. McNamara's job was not finished until weaker rock was removed and solid rock exposed on which concrete could be poured. California agreed neither on there being as much rock involved as McNamara claimed nor on the price that should be paid out.

Although months of back-and-forth occurred, there was eventually movement on the California side. Ultimately, owing to evidence presented that Golze had admitted, in writing, to one of his senior staff members, after the report of an independent geotechnical consultant, that he believed the change in conditions was warranted, the court case ended in favour of McNamara, in the millions of dollars. The most burdensome job in John's California tenure was moved out of the losing money column.

While he was back in Toronto for only a few weeks during this time, McNamara was making promotions as part of a restructuring plan to facilitate growth in business and expansion in capabilities. John was

The Bahen family, while visiting Expo '67, at the Montreal restaurant Au Lutin qui bouffe, which was famous for its piglet. John (left), Susan, Margaret, Michael (holding bottle), Stuart and John and Margaret's niece Cathy Zimmerman.

promoted to Vice President and General Manager Heavy Construction, the news being announced in the *Toronto Telegram* on Tuesday, September 1, 1964. The letters of congratulations that he received highlight, once more, the powerful support of mentors such as Bill Tate, who wrote a personal note to him, as did many in John's extensive network of business allies in construction, materials, equipment and consulting disciplines. He was, after all, still only thirty-seven years old. Common among the notes was not just their subject but also the light-hearted humour with which the writers teased John about his education, his receding hairline and his weak golf swing. The friendly, sarcastic tones were balanced well with the more intimate ones that reminded John that his mother would have been proud and that he worked harder than most to get to vice president.

It was an exciting time for John and Margaret. The two older Bahen children were in school and the youngest, Michael, had started nursery school. Soon, in early 1965, the family moved to Foursome Crescent in Willowdale in order to have more space and be closer to a school with special needs teaching for Michael. They also had a small cottage at Nottawaga Beach on Georgian Bay, purchased about five years earlier, in part with the help of a loan arranged by Harold McNamara.

9

KIEWIT ENCOUNTERS

"WE seemed to be bumping into each other all the way along." So says John about his encounters with Kiewit on his projects over the years. It seemed fated that he would work with the company one day. "They were building all the great projects, and I wanted to be with them."

Kiewit's work in Canada or near its borders tended to mirror John's early career moves. While bidding on work on the west coast, Peter Kiewit had personally taken an interest in the Toronto subway project during the same years that John was working his first job as a summer student for the TTC on the underground survey crew. That was the first time he heard the Kiewit name.

The Dominion of Canada Corporation was established by Kiewit (1949) in partnership with Arthur Johnson Corporation, a New York City construction company with a great deal of subway experience. They took on Toronto subway work by early 1950. Kiewit would ultimately purchase the Arthur Johnson Corporation, working under the name of Johnson-Kiewit Subway Corporation.

Kiewit was doing TTC subway work when John decided to leave his summer job and hitchhike to the Turnpike in 1950. Kiewit also had Turnpike work. When John was in Goose Bay, Kiewit did substantial work with the U.S. Army Corps of Engineers at Thule, Greenland, the other base critical to Operation Bluejay. Kiewit was also involved at Baffin Island, building navigation stations for the Distant Early Warning

(DEW) system in the north. This may have been his first access to real information about the company, although, at the time, he was more concerned with achieving the expectations of his American military bosses than prospecting for future career opportunities. These would materialize fourteen years later.

After John was hired by McNamara and at work in eastern Ontario on jobs directly or indirectly related to the St. Lawrence Seaway, Kiewit was working in joint ventures on the American portion of the Long Sault Dam, its canal and diversion cut, the Iroquois Dam and a major powerhouse. Although he doesn't recall meeting anyone from Kiewit then, it was another reminder of the company's consistent presence on major work, the kind he yearned to be part of.

Most people in the heavy construction world knew Kiewit by reputation, not only because of the challenging and big-money jobs it was completing but also because so many contractors were consistently losing work to it and shaking their collective heads at how the company managed to be the lowest bidder so often. Another factor that got people talking about the company was its unique structure, which incorporated employee ownership. Though not yet the construction behemoth north or south of the border that it is now, Kiewit was establishing its pattern of dominating large-scale public infrastructure projects.

John saw Bob Wilson, then Kiewit's vice president, at the Churchill Manitoba airport when Wilson was touring a number of military construction contracts being executed by Kiewit's competitors. The two had a lengthy conversation covering the outlook of the industry, McNamara's reach to international work and John's level of job satisfaction. It was not long after the Churchill meeting that Wilson contacted John and introduced the idea of a Kiewit-McNamara joint venture. Although not a large contract – a sectional tunnel under the St. Lawrence River on the east side of Montreal – it was comparable to work that Kiewit had completed in San Francisco and not something that McNamara had the expertise to bid on alone. The opportunity for Kiewit to land a contract in Quebec was inviting to the company at the time.

"It was 1963 when the bid for the Boucherville Tunnel was assembled," explains John. "The project involved tunnel sections being floated into place and carefully ballasted down into a prepared trench. This was a Kiewit specialty and, although McNamara had tunneling experience in various locations, the joint venture was a mutually beneficial one."

In the process of preparing the joint bid, John and Sandy MacDonald met a couple of times in Montreal with Bob Wilson and Martin Kelley, an engineer-estimating specialist then working with the Home Office Engineering Department. When the final bid review process began, John had his first experience with the two-estimate process for which Kiewit is well known. He and MacDonald were asked to come to Omaha for the two to four weeks' time it would take to compare their estimate with one independently prepared by another section of corporate Kiewit. John was told that several people were interested in meeting him, but he was baffled, at the time, as to why.

John was very impressed with the Kiewit process, from the excellent level of co-operation and the composition of the bidding team to its collective knowledge of the materials, equipment, employment situation and location. It reflected flawless estimating, record-keeping and contract cross-referencing, areas that John had grown weary of wishing to improve at McNamara. He thought that the methodology was miles ahead of anything he had seen.

The bid was unsuccessful, but the relationship between John and Kiewit was forged. In the pre-submission weeks, he saw the top executives of Kiewit almost daily, including Peter Kiewit, who "came by to see how the Canadians were doing." Before the few weeks were up, the idea of Kiewit establishing a Canadian office was raised. Less than a year later, living in San Francisco while working on the Oroville Powerhouse job, John saw Bob Wilson from time to time at the hotel where they both stayed. John's early impression of him was that "he was tough, with a smile. He was a decisive guy." Peter Kiewit Sons Inc. had a job site not far from Oroville, bringing Wilson to visit fairly often. Wilson was headed out to play golf one Sunday morning, when he and John exchanged niceties and had breakfast together.

"He was going golfing," John recalls, "and I was going to one of our jobs that he knew about. It was a rock excavating job where we were taking the rock out with scrapers. This was a little unheard of at that time, but the scrapers were specially made for this. I guess my going to work and his golfing on a Sunday afternoon – the contrast kind of grabbed him."

Wilson asked why John had to keep working, half-teasing that even Kiewit took Sundays off (if only Sundays). John took pains to explain that the job was fraught with challenges and that they were working around the clock with scrapers. In fact, because of the pressure of the Oroville job and having to work in conditions they could not have anticipated, John was working every Sunday. Undeterred by the explanation, Wilson asked why he didn't have any time for golf.

Wilson was a hard-working man himself, but the fact that he was able to golf on Sundays while John had to return to the stressful management of a dangerous, money-losing California job made John realize that he might be suffering more than was necessary. It was around this time that he made the quiet decision that it would not be long before he made a career change.

Despite John's encounters with Kiewit, it shouldn't be concluded that it was just a matter of his being in the right place at the right time in late 1965 when Kiewit started to seriously pursue him for its Canadian expansion. With him, it is always a risk to underplay the influence of his strategic thinking and meticulous planning. In fact, things were working out exactly the way he wanted.

In John's mind there had been ample evidence of weak business decisions at McNamara, but he believed that there were also hints of impropriety around the place. A general fear was developing among staff that they might need to worry about their jobs, not to mention getting paid. Eighteen months of being back and forth between Toronto and California, topped with a shooting, a state-level law suit and rumours of misappropriated funds helped convince John to consider his options. Ultimately sensing that insolvency might be the outcome of family infighting, he came up with a short list of career options. His high esteem for Bob Wilson, whom he referred to as "an outstanding man and an outstanding builder," helped put Kiewit high on that list. It was Wilson's compelling personality and work ethic that inspired John, who thought that Wilson possessed knowledge and skill that went far beyond engineering and construction. Wilson was another worthy mentor, and their relationship became not unlike that between John and Bill Tate.

By the time the top three executives of Peter Kiewit Sons Inc. began to seriously offer John options to join the company, he had come to know how successful Kiewit was, and he knew its work and some of its people. But, perhaps most intriguing to John, was its desire to expand its Canadian presence. Active in western Canada and intending to seek work in other areas of the Canadian market, Kiewit provided an opportunity for John to carve a niche for himself with a company that he could be proud to work for. This provided the solution to his desire for change. Kiewit had been in the forefront of his mind for a couple of years. He had already discussed with Wilson the corporate structure, Peter Kiewit's vision for business and growth and the stock program.

Bob Wilson called John and put it plainly: "John, I'd like to talk to you about coming to work with us."

10

WEIGHING HIS OPTIONS

It is not characteristic of a typical engineer, and certainly not of John Bahen, to take action without analyzing all of the repercussions and possible outcomes. But his need to leave a company that he felt had serious shortcomings was greater than his need to have the next step meticulously planned. Although he had spoken with Bob Wilson and had met Peter Kiewit and Walter Scott, Jr., then regional manager for operations in the Great Lakes area, no details were on the table when John resigned his position with McNamara. The following months may have been characterized by uncertainty, but he had many options to consider.

At this time, he created Bahen Engineering, although it was a short-lived endeavour. John realized that he would not want to be away from job sites and the field for long, but he was confident there was plenty of work for his estimating skills. He acted on behalf of Public Works Canada in Saint Maarten, where the federal government was supporting construction of an airport on the Dutch-governed side of the island. The Canadian government did not want the construction project to take place without some oversight by an engineer of its own, so John lived in Half Moon Bay when he was needed on-site to confirm estimates and double-check costs.

At the same time, he entered into discussions and went as far as preparing a draft offer to purchase a construction company in eastern Ontario. He wondered if he was ready to run his own company. It was not overly disappointing to him when the timing of the purchase dragged on, and John turned his attention to his other options.

Interestingly enough, one of his job offers came from McNamara's largest local competitor, C.A. Pitts. Sydney Cooper, who led Pitts, was a business adversary, but he clearly recognized John's talent. Years later, when John was retiring from Kiewit, Cooper spoke about his desire to have John on board at Pitts, remarking that he would rather have Bahen with him than out there in the market working against him.

John went to his closest friend, Denis Evans, who, together with his partner Roy Steed, was active in the construction business (Steed and Evans) in southern Ontario and in the northern United States. John asked Denis to review his options with him. These included offers such as working in the hydroelectric power field as a corporate engineer, some less formal opportunities to take upper management positions at other construction companies, as well as an offer from Peter Kiewit Sons Inc. Evans says that he knew the contractor, builder and engineer in John would not be satisfied with the hydro placement, although there was prestige in it, and he said so.

Undoubtedly John Bahen could have made a success of anything he chose. But accepting Evans' advice, the engineer and builder in him went with his gut and decided to stay in heavy construction rather than be diverted by a lucrative position of authority. He remained true to a single feeling of accomplishment, the moment when he looks upon a completed project and can say, "We built that." John has never settled for less than what he wanted. "He is a builder," says Evans, "and he has done so many remarkable things since making that one decision."

John remembers well the conversation he had with Pete Kiewit. "Pete gave me my first interview – I should say my last interview."

They discussed what John had to invest in the new Canadian endeavour. At the time, Kiewit's president pondered replicating a situation that he had established with Toronto commercial builder V.K.

Mason. Kiewit had brought the company under its mammoth corporate umbrella and directed commercial buildings work toward the Mason operation while holding the entire performance bond. The tremendous financial success of the alliance with V.K. Mason was precedent enough for Kiewit to want a similar connection.

"They wanted to do buildings, like First Canadian Place, but I'm not a buildings man," explains John.

Pete asked John how much money he thought he would need to start up his own operation – a Bahen-run and -centred operation under the wing of the Kiewit corporation. He calculated in his head what equipment would be required to bid roadwork around the Toronto area at that time and said that it would take $60 million to $70 million. In jest, he then pulled a wad of small bills from his pocket and grinned.

Peter Kiewit replied, "Oh, I guess you don't have enough then! You might be right about those numbers, though. Well, let's talk about it another day in the future."

John notes, "Pete could give the most inspiring speech. He was a marvellous man. A great leader. He was an inspiration to everyone who worked for him. But I wasn't about to lead my own company. I had far too much to learn and still just wanted to build." Having his name on the banner of a company had never been important to him.

John travelled to corporate headquarters in Omaha, Nebraska, to meet with Kiewit, Wilson and Scott, Jr., to develop a detailed plan for how Kiewit could access the Canadian market and what role they could see for him in the Canadian expansion. The eventual tangible employment offer put the onus back on him, a challenge he rejoiced in. "Why don't you see what work you can find in Canada?" was his directive from Bob Wilson. That was really all John needed to begin. He felt confident the details would fall into place. When asked what Kiewit put on the table to lure him over, he responds with a chuckle: "Lots of work, some new challenges, not much money."

Kiewit was not offering to even match the salary that he earned at McNamara, or any other offer for that matter. Kiewit, John learned, was not known for its salaries. The November-December 1972 edition of *KIE-WAYS* magazine articulates the draw that Peter Kiewit Sons Inc. had for him. It explains that good wages are not enough to draw a man to take on taxing work and leave his family for weeks at a time; instead the drive comes from "the strong desire to be a part in building something useful." A member of the board of directors writes that, for an engineer, it means having pride in seeing plans executed in a "smooth sequence of operations."

A worker-satisfaction-focused management system, tied to productivity and construction excellence, was something that John identified with, and finding a fit with Peter Kiewit Sons Ltd. was as exciting to him as his first job.

What really clinched the deal for John was Kiewit's employee-ownership model. This was the opportunity to augment the mediocre starting salary, and John's competitive instinct thrived on the notion that individuals were given annual merit-based opportunities to purchase corporate stock. It was basic and brilliant to him: if a person's work benefitted the company in a financial way, the opportunity to purchase a certain amount of stock (with financial assistance from the company when necessary) was offered. This translated into a feeling that he was working for himself but, at the same time, was protected by a conglomerate.

"Ownership. Stock ownership. That I really liked," John says. "That could make up for the difference in salary. I wanted to make more, and I wanted to work harder too. Kiewit was a place I knew I could do both."

And he did well by it over the years.

Perhaps what John admired most about Peter Kiewit personally and, by extension, about the company, was that the corporate emphasis was on the truly important details of construction. "The profit wasn't the most important part, but it resulted from focusing on the most important part of the business, which is getting work at the right price. The company's success, which came from the philosophy of a brilliant man, was to make sure that the job being bid was analyzed properly by the people who were likely going to build it."

11

THE WELLAND

THE first location of Kiewit's Toronto bidding office was a space above a lumberyard in the city's west end. The understated, humble, cost-conscious approach of Kiewit and of John Bahen showed in its practicality and simplicity. The office space was affordable (former employees say "cheap") and serviceable rather than expensive and showy. John never wanted a supplier or customer to come in to an ostentatious office because he felt that would convey the wrong impression about Kiewit.

Work began immediately. John prepared bids for contracts that he had scouted when he was challenged by the Omaha executives to get work. The Welland Canal By-pass project, a by-pass of the "Fourth Canal," was required to improve a winding and narrow channel route in the section that bisected the town of Welland. The by-pass would straighten and widen the canal and, by going around the town, correct traffic problems created when vehicles had to cross the canal by bridge while it was in operation. With proven talent from Kiewit's work on British Columbia's W.C. Bennett Dam at his disposal, the bidding was successful, and John hired new staff in Toronto, many of whom launched their careers with Kiewit. Realignment work between Port Robinson and Port Colborne began in 1967 and continued for almost seven years. Kiewit's Toronto operation took more than half the contracts on the Welland By-pass project, establishing the Kiewit name, reputation and quality of work in the Ontario construction market. It also established John Bahen's value to Kiewit.

It was a time when John began to understand exactly how sharp each item in an estimate had to be. Kiewit faced solid competition from C.A. Pitts and Syd Cooper. Pitts also worked at low cost, but Kiewit won out on its ability to draw on a pool of equipment and people made available from other districts. While the competition rarely got the work that Kiewit wanted, it made it even more interesting at the Welland.

The Welland Ship Canal is located between Lake Erie and Lake Ontario, achieving a level change of over three hundred feet and enabling large ships a passage that avoids the rapids on the Niagara River. There had been four phases of building in the Welland Canal's history, beginning in 1824, with each expansion responding to demands of the shipping industry and changing ship dimensions. The Fourth Canal, a meandering nine miles, was constructed starting in 1918. Traffic congestion in the town of Welland prompted town officials to lobby the federal government to approve a relocation of the Fourth Canal to the east of the Town at an estimated cost of $107 million.

The scope of this work involved relocating 8.3 miles of canal, creating a new underpass for highway and rail tunnels and excavating a total of over sixty-five million cubic yards to increase the canal width to three hundred fifty feet and depth to thirty feet. The section that passed through the Town of Welland had numerous bends and six bridges to be navigated, adding to the visual obstructions and narrowness that already plagued ships' captains and caused time-consuming delays. A public ceremony for the project was held for the first shovel in the ground in June 1967. By then, Kiewit had mobilized on its first contract. The excavation work was broken into ten major contracts, of which three, won by Kiewit, consumed the bulk of Kiewit's time at Welland.

It was perhaps these Welland contracts that first showed John the benefits of working with a massive company where expertise on very detailed work could be found simply by asking. "No matter what you needed," he says, "or what job you took on, there was always someone at Kiewit who had done something like it. I made use of that."

The innovative methods used for excavation at Welland, based on advice and precedent of work performed by Kiewit elsewhere, were a key contribution to bidding the work at the lowest price. John was eager to hear the advice of his own and Kiewit's specialists, as the work required multiple opinions and a creative solution.

"Our experience is that this is a dragline job, and Pitts' experience was that this was a scraper job. All the marine contractors in the canal area bid this job with their dredges. They couldn't compete. To do it in the dry and not have a sloppy mess for a mile around there – they couldn't compete with our method."

The choice to use different machinery from the competition meant that Kiewit got a majority of contracts at Welland because their price was at least 20 per cent less.

John's focus was to lower costs, improve profits and, just as important, become more competitive. All of these factors led to developing more opportunities. Welland was no different. At the end of each day, no one went home without knowing his best hours of the day and the costs the project was running. This included dragline operators, who were quite motivated by having their numbers posted and knowing those of co-workers. Competition is alive and well in heavy construction, and the project benefitted at Welland.

Drawing on his experience on the Mangla Dam bid in Pakistan, John appreciated the value of having motivated and steadily challenged employees. He was impressed, as well, by the significant impact that construction process innovations could have on the bottom line. Even within a contract, costs could be drawn down by affecting time, quantities, machinery and men. He became an expert at this.

John was quick to recognize talent, and many engineers and superintendents who came to Kiewit during the Welland days became trusted staff on his team. A good number found a long career with Kiewit, and still others used it as a proving ground to expand their potential. Geoff Lindup was a young, hard-working British engineer looking for work to stay in Canada when he looked up John Bahen at Welland in 1968. He sent a letter to the Kiewit company on the recommendation of a former colleague of John's at McNamara, was hired and went straight to Welland the same year, while the first contract was being built. Lindup spent a lot of time, in those first weeks, walking around with a stopwatch and a pad of paper: his induction into the detailed and precise science of estimating and contract management at Kiewit. He got to know dragline and dump truck cycling times well. It may not have been clear, at the time, how critical the time studies, precise recording and analysis of dragline cycles were to the operation, but it was the Kiewit way, and John was committed to it on his jobs. Attention to detail and intricate knowledge of line items are keys to bidding work at the right price, but also – and this is where John excelled – to seizing every available opportunity to improve on performance or lower costs during construction. Everything was recorded and then cross-referenced with the prices bid and hours forecasted.

Only a small collective of engineering grads excel in a place where business skills marry engineering knowledge. But combined, the two skill areas produce success from the numbers.

Reflecting back on their careers, Lindup remarks, "John would be more the businessman than the engineer. He is not a technician; he understands engineering principles absolutely – primarily soils and material – which played a big part in our business. But as far as doing rocket science calculations, neither he nor I can do that. We hired people to do that."

Sometimes technical innovation predicted increased profit, but often enough just a long-shot idea would do it too. On one of the contracts, prior to bid, a novel idea was presented to John. Instead of bringing in quarried fill, why not use the slag sitting in great heaps not far away at the Algoma Steel plant? The contract required placement of liner material on the excavated surfaces of the canal. John spoke to Algoma about the slag purchase and to the Seaway Authority representatives about the acceptability of using the slag for filter. Liner material was required on the slopes of the canal where wave action and propellers from passing boats erode the slope and stir up silt.

Kiewit bid the job on the premise of using quarried and processed rock as the base or filter layer with a coarse rock material on top (together these two materials would constitute the liner system).

Says John, "This is a place where we made a million dollars just by keeping our eyes open. The slag that was available locally was quite acceptable as a liner material."

He is proud still of these first creative shortcuts to being the lowest bidder, making a profit and accessing more work.

12

PROJECT OF THE CENTURY

JOHN'S cheerfulness regardless of what business challenges he was facing was the feature that most people noticed immediately, second to his work ethic and perhaps third to his incredible ambition.

"You could always get John to laugh," says Geoff Lindup. "Even though he got much more serious under the pressure of James Bay, he was still John."

Maintaining the ability to enjoy the people he worked with was important to John, and finding a way to laugh helped him avoid becoming too mired in the problems of the construction business. It was part of his style. As he says, "You cannot think about the risk all the time or it will get to you; others around you will see it and think they should be worrying too."

The nature of his role changed with the increasing responsibility. He was overseeing more people, and the risks taken on the job were more significant. He felt the pressure of the individuals and families whose future comfort relied on his solid decisions. But he remained the one who could laugh at himself, and, although often obstinate and extremely confident in his authority, he was always able to return immediately to the man his staff expected to be both their leader and their champion.

As a manager, he was always interested in the details, anxious to visit superintendents on jobs every couple of weeks. On these site visits, he had a way of asking questions and pointing out improvements and corrections that made work for his staff. As such, in addition to the nicknames "Dad," "The Boss" or just, "Bahen," he gained a new nickname, "The Big White Bird." In other words, one who flies in once in a while, shits all over everything and flies off, leaving the clean-up to the lesser birds. One practical description by one of the young engineers was that, "He came and gave his support and said, 'Get it done,' and then he would come in next time and ask, 'Why isn't it done?'" John finds this as funny (and accurate) today as he did then.

The James Bay Hydroelectric Project was a political strategy long before it was a construction project, not unlike other long-discussed and -debated infrastructure undertakings (such as the New Jersey Turnpike). It was a promise attached to a Quebec provincial leadership and regime change, intended to be a motivation for the people, in general, and a stimulator of voter support, in particular. It was also an engineering marvel of unprecedented scale. Called one of the five largest contemporary river-based projects in the world, its ultimate capacity (with twelve dams) was within the range of twenty-five thousand to thirty thousand megawatts of hydro. It was the kind of work that John Bahen craved. It was large enough for his company to take contracts gradually and consistently until Kiewit's status was more broadly established. It would wedge Kiewit firmly in the eastern Canada market and, because of the project's scale, spread its reputation to the whole of the country. He knew it was the path to achieving his promise to Pete Kiewit to establish a Canadian arm of the parent company.

John and his team, Gene Bednarski, Bruce Daniels and Dave Callander, were exceptionally skilled at project intelligence. In the case of James Bay, they needed to use every method possible to assess the risk of pursuing work, the future of which would be decided by the unpredictable Québécois electorate. It is said that, if you are reading about a job, you have likely already missed a chance to get it. So they listened, asked questions and stayed informed. In order to be on top of what construction work was coming, John had his team foster productive and congenial relationships with other contractors, heavy equipment manufacturers and union representatives who would also be benefitting from James Bay's "go" status.

Running a district at Kiewit meant operating an independent profit centre, and although the resources of the company were there for the asking, each district had its own equipment and people and managed its own work. Each district manager was responsible for the business of his district and for having enough work to use machinery and people optimally.

The Executive Committee – the top executive management of Kiewit's headquarters in Omaha – was always involved in decision making on what large jobs would be bid, and at what price. Pete Kiewit personally had the authority to veto any project being considered by a district manager.

Long before estimating for James Bay began, John had the forethought to establish a Quebec business presence in the form of a Montreal-based office with a French name, Les Enterprises Kiewit Limitée. It incorporated in 1969, a subsidiary of Peter Kiewit Sons Co. Ltd. (the Canadian corporate name), with the sole purpose of engaging in work in Quebec. This was a tactically brilliant move, which became obvious when the first bid documents were opened and Kiewit was the lowest bidder. Any controversy over an American company being the first to break ground on Quebec's legacy project was muted by the company's pre-existing Quebec presence.

John was under no illusions; he continued to work with full acceptance of the construction industry norm that no more than one bid in ten was successful and so chose to focus his attention on what the timing of the award would be. That was typical of his positive attitude: to accept the reality of things that he could not change and instead pour his substantial energy into available opportunities. If more than one bid was a winner and if more than one winner was announced close to another, the stretch in his resources might eliminate his potential profit because it could require purchasing new machinery or training a freshly hired workforce.

As is typical of a massive public sector work in which the client/owner is government or an arm's length subsidiary, information was released slowly and in a managed, politically advantageous manner. No one seems to recall who discovered that James Bay was going to be publicly announced and when the first bid deadlines would be, but John was so sure that it was a golden opportunity that he put people on it months in advance. Well informed by staff of the Montreal office that Robert Bourassa's Liberal government was expected to be elected, and knowing that Bourassa campaigned on economic prosperity for Quebec, the cornerstone of which was the James Bay proposal, John prepared. Even while construction efforts were deeply committed to the Welland Canal, and with concerns from Omaha that the James Bay contracts were a capability stretch for the Eastern Canada District, John championed the bid preparation relentlessly.

Unlike the Turnpike, which had been a foregone conclusion, James Bay was at risk of elimination entirely. Its future depended on the provincial election. If Bourassa failed to win a majority and be elected to the legislature of the Province of Quebec, it was highly likely that the James Bay project would be permanently shelved by his critical opponent, who had promised to do so. The risk that John was taking, banking his district's resources almost solely on bidding and building work for James Bay might have ruined his career at Kiewit, or worse, his long-standing reputation for having an eye for great work. But he did not hesitate.

More than once he went to Omaha to describe to Pete Kiewit the scale of the initial access roadworks and construction programs as a whole, as well as his plan for getting highly profitable work using joint ventures and partnerships (at first) to dilute risk and build credibility with the client. He felt strongly that the roadworks contracts would open doors for Kiewit to gain valuable operational and costing experience, build a team and ultimately qualify (and be in a position to get work at the right price) for larger and more lucrative bids to come.

While there had been many doubters (including many within Kiewit's Omaha management group), "The Project of the Century," so billed by the new premier,* was online and launched by April 1971. Premier-elect Bourassa declared that the project would proceed, and, to prove it, created a government-owned corporation – Société de développement de la Baie James (SDBJ) – to oversee development of the James Bay territory in the areas of forestry, mining and tourism. An energy corporation – Société d'énergie de la Baie James (SEBJ) – was also created to oversee the development of le complexe électrique de la Baie James (the James Bay Hydroelectric Project).

Phase I of the hydroelectric development would be composed of dams, dykes powerhouses and other structures along La Grande River and its tributaries. The James Bay territory and neighbouring Labrador were a potential combined power resource of possibly over thirty thousand megawatts. Phase I focused on La Grande, one of at least five Quebec rivers, the capacity and geography of which make them major hydro power producers. This phase involved the construction of three powerhouses and the diversion of two rivers to the south and a river to the east. The total installed capacity of these generating stations was over 10,000 megawatts. The cost of Phase I was approximately $18 billion. Ironically, while promising eighty-five thousand person years of jobs to Quebec workers constructing six miles of dams and thirty-four miles of dykes, Phase I of James Bay had neither an airport nor an access road when announced. Route 109 went to the town of Matagami and no farther. Access roads were needed to penetrate forest hundreds of miles farther north before a single bulldozer would get anywhere near La Grande River.

---

* Announcement by Robert Bourassa of the commencement of the James Bay Hydroelectric Project, made in a speech given April 30, 1971, to five thousand Liberal party members at the Petit Colisée in Quebec City, Quebec.

This was the scale of work that John knew would establish and sustain a Kiewit presence in eastern Canada for the duration of his Kiewit career. It was his dream, and it would be his crowning achievement. He knew that construction of the first phase of Le complexe La Grande could take ten to twelve years.

Walter Scott, Jr., says: "When we talked in Omaha about John, I think most of what we talked about was his legacy. That was the first major work that was done up in Quebec and bigger than anything done in eastern Canada. The fact that John took it on and that it continued over a whole series of jobs helped us hire and build the group of people that continue today. They are working for the company and doing good things just like John taught them."

John at the crusher operation, Mile 14 of the Matagami road job.

13

FROM ROADS TO DAMS AT JAMES BAY

WRITTEN descriptions of the landscape surrounding James Bay, and in fact writing in general about the Canadian Shield and northern expanses of Canada, laud the dramatic greens of glacial lakes, the magnificence of ancient rock formations and the pristine beauty of untouched old growth forests. Such adjective-laden text would have been lost on those who made the trip to James Bay by plane from Montreal to Matagami and then still farther north by truck or bus to the road camp at Mile 48. The Kiewit men and women endured harsh conditions. Temperatures were regularly -13° to -22°F, and there was no settlement with urban conveniences for hundreds of miles from where the bulk of the construction was taking place.

The amount of running water in James Bay is due more to the way the frigid temperatures eliminate the possibility of evaporation than to geography. It is virtually uninhabitable, yet, for more than twenty years, the people of Kiewit's Eastern Canada District sent willing teams of people to James Bay to be part of the largest construction project in Canadian history, building through the winter, enduring temporary encampments and weeks of isolation.

One third of all fresh water flowing in Canada's rivers flows into James Bay and Hudson's Bay. There is almost no comparable construction project of this scale in North America. "The James Bay project is one of the largest energy projects ever. It is no mere megaproject. 'Mega' denotes only a million, but the magnitude of this project is measured in billions of watts and dollars. 'Giga' denotes a billion. This is a giga-project."*

Given the developing media profile, the first contract bid opening on December 6, 1971, was a public event, televised and broadcast on radio in Quebec. Excitement for the economic benefits of the giga-project construction to Quebec contractors and builders was at its peak before the first contract was even let. The Bourassa government had been elected on the promise of economic vitality and growth for the people of Quebec, and expectations were high.

When the bid opening showed Quebec contractor Simard-Beaudry to be the lowest bidder on one road contract, only to be immediately disqualified because of a rather simple error that did not respect a technical clause, disappointment was palpable. This meant that Les Enterprises Kiewit Ltée was the lowest bidder of the eight. Despite its official French status in Quebec, Kiewit was continually described by media as a Montreal subsidiary of a massive American company. The newspapers ran reports attacking the project, the government and Bourassa himself, arguing that the detrimental impacts to Quebec had begun already, with the first bid opening, because a Quebec firm was not being protected by the process. Minutes after the Kiewit bid was

---

* McCutcheon, Sean. *Electric Rivers: The Story of the James Bay Project*. Montreal: Black Rose Books, 1991., p. 8.

reviewed, found without errors and declared the lowest bidder, the site office phone rang at the Welland Canal, where René Plourde, who was French-speaking, took the call. Of this turn of events, which brought the Eastern Canada District into James Bay, John recalls: "It's just the way it worked out. We ended up with two jobs. We really didn't want the second one [a road contract]. It was a little too much for our equipment and our people. We bid on two jobs, and one of those jobs was bid by the other guy who couldn't take it. I'd never seen any of their top people, but we would bump into them all the time soon enough."

The value of the two roadwork contracts was $20,965,530 and involved construction from Mile 22 to Mile 88.1 of the access road from Matagami to La Grande River. The road would be the sole access to La Grande Complex, with a completed length of 411 miles. John says he never expected these contracts to yield more than a reasonable margin of profit. But the strategy was about more than getting the job; it was about getting significant work in Quebec and a toehold in the hydro project. During the challenging construction period that followed, he would find himself defending that narrow margin, in fact, and struggling to maintain it. But his strategy was sound and proved brilliant in the decades that followed.

The conditions in sub-Arctic northern Quebec in the winter months of 1972 were completely unlike anything that the Kiewit team had ever experienced. All equipment, materials and people had to be trucked in on winter roads and over temporary ice bridges, typically from Montreal, during all kinds of weather. Keeping access open through all seasons was critical to ensuring that jobs would run on time. Ice bridge construction was the highest-risk work of all. Rivers were fast flowing, and methods were being invented as the crews worked. At the Waswanipi River, the ice was reinforced with logs that were trucked in, laid onto the three- or four-foot-thick ice and flooded by pumps circulating water continuously from beneath the ice cap up and over the surface, while the river flowed wildly underneath.

Not only did severe climatic conditions have to be contended with, but there was also incredible pressure to build quickly. John knew that jobs were subject to very critical timing expectations because, by spring thaw, the equipment, fuel tanks and a full camp had to be in place. The Kiewit miles on the Matagami Road (later named the James Bay Road) were early in the trip north (starting at Mile 22) and therefore essential to getting living arrangements set up farther to the north for hydro work. Every contract for every major structure in the Complex required use of that road: the Kiewit miles needed to be passable for every other contractor building farther along. Timing had to be co-ordinated. Kiewit committed to a schedule of two shifts a day, six days a week. By May 1972, when the thaw came, a substantial camp was constructed at Mile 48 that had trailers for housing, shops, offices and a warehouse. A portion of the road was constructed to a standard suitable to function as an airstrip in anticipation of the summer construction season. Even with a more substantial camp constructed, hours were lost each day to transporting personnel by pickup truck and bus over partially constructed, swampy and muddy tracks. These costs were not anticipated in advance of arrival in the north. There was a substantial risk of profit loss due to a lack of experience with these elements and comparable cost estimates.

"Our first year at James Bay was a very tough year," John recalls. "No matter how much time I applied myself to the problems, it was hard to see that we were making tangible progress on any front."

John Loewen had come from work on the Welland Canal to manage the road jobs at Matagami. The key staff who formed a team for these contracts, many of whom had been with John Bahen since the early Welland Canal project days, remained a presence for many years and in future contracts in James Bay, rotating through or rising in various roles of increasing responsibility. Key team members included John Smith, Bob Cranney, Jerry Johnson, Ken Austin, Emile Merceau, Bill Wietzl, Emil Gaspar, Ron McGuire and Ralph Dickson.

At Matagami, a young civil engineer, Donald Quane, assumed the role of field engineer (and would be promoted to project engineer only three months later). John had interviewed him at the University of Montreal and offered the job at James Bay right after a second interview. Two days after that interview, Quane was getting on a plane to assume responsibilities far beyond his expectations. Remarkably, the complete length of Kiewit roadworks was completed in months to a standard that allowed vehicles to use it safely. The part winter/part all-season road opened February 1, 1973, with significant winter-only components. The all-season road opened in October. By 1976, the entire road had been paved.

When Walter Scott came to visit one time, as he and John were crossing a bridge, Scott spotted a canoe in which a man wearing a Kiewit hat was being paddled by a Cree woman (with whom, it turned out, he was living) and announced that was the first time he'd ever seen someone come to work that way. In fact, this gentleman, who lived to fish, hunt and have children, happily stayed after the Kiewit team had all left the job. John found that very romantic.

~

During his time at James Bay, John committed himself to nightly calls home, speaking with each of the children. As was his style, he had nicknames for each of them. Susan was Mimi because, as a youngster, she couldn't pronounce her own name and thus referred to herself as "Me! Me!" Stuart was called Bucky, and Michael was Gavi or Gavalotzky. John and Margaret even christened three of their boats, over the years, with the children's nicknames. As much as possible, he spent Saturday afternoons and Sundays with the children (when he was in town), made it home for the major events in their lives and tried for more frequent long weekends at Nottawaga Beach in the summer. They had also moved to 7 York Ridge Road, which they would call home for the next twenty-seven years. John had always wanted a charming stone house and found it on York Ridge.

Of course, Stuart's good friend, Bruce Randle, who was almost part of the Bahen family, had to first check it out and give his approval. Stuart and Susan recall their dad being there during these family times, and, with the schedules of their school lessons, sports and social lives in Toronto, they were not really bothered by his absences. They knew where he was and studied maps with him on weekends to understand how far away Kiewit was building. They attended high school at York Mills Collegiate in Toronto, and Michael, who had distinct learning needs, was enrolled at the Phelps Academy in Pennsylvania.

A few years into the James Bay contracts, John had confidence in the supervisors he had in place and took more of a management role out of the Toronto office. This made him more of a regular at the family dinner table through the week. Susan remembers vividly the change in the family structure later in the 1970s. With Stuart starting university and Michael still at the Phelps Academy, teenaged Susan and her father had a special, close relationship together over many quiet meals. Stuart studied civil engineering at Queen's University and worked summers at James Bay. Once he graduated, he joined Peter Kiewit Sons Ltd., ending up in the San Francisco area and remaining with them for many years.

The Mimi, named after Susan's nickname.
The Bahen home for twenty-seven years, at 7 York Ridge Road, Toronto.

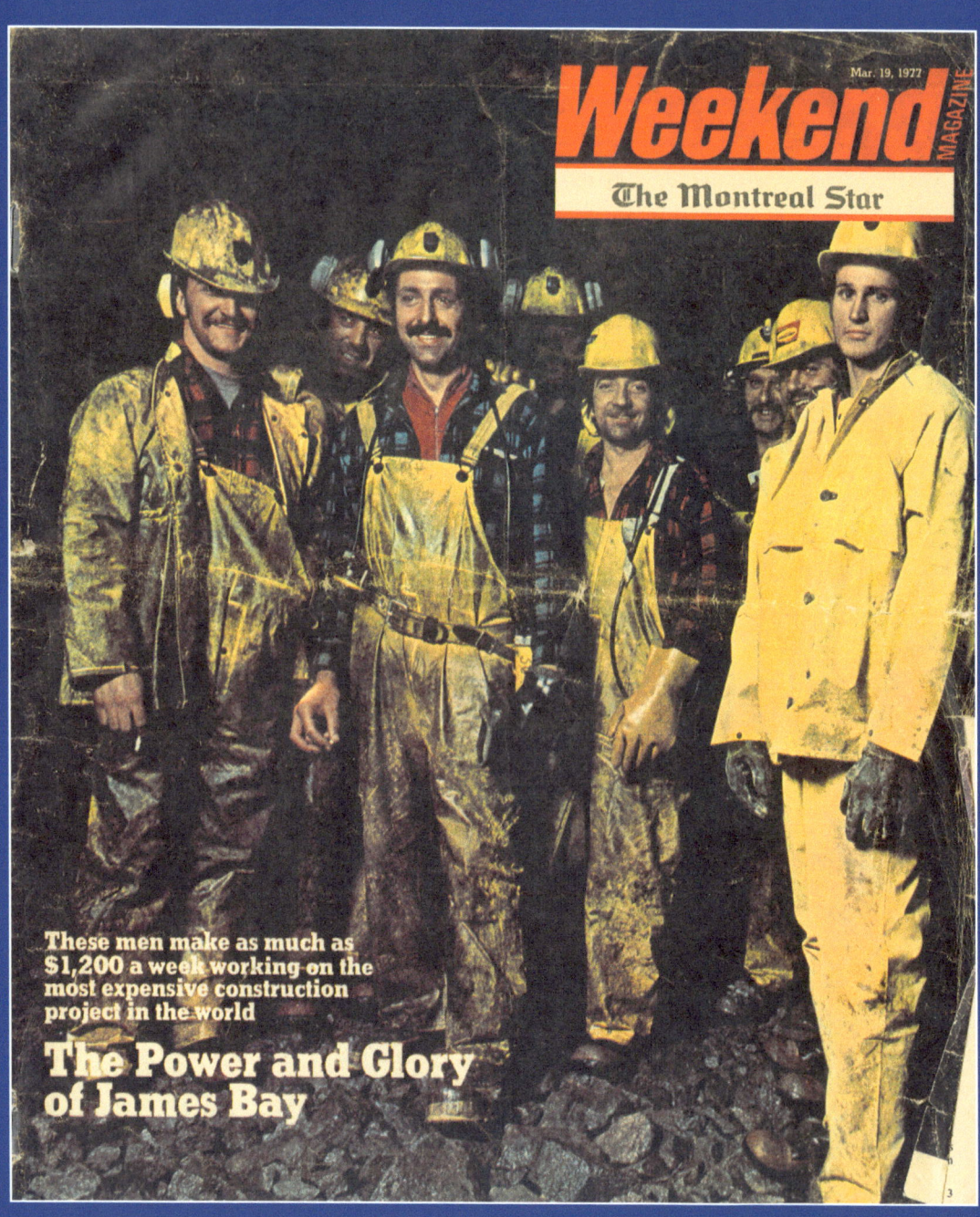

The article that inspired Ted Chant to join Kiewit's James Bay team.

14

BUMPS IN THE ROAD

Given the profile of the project and the enormous amount of money being promised for local employment, perhaps it should not have been shocking to most that some who sought to prosper would use any kind of leverage for access to James Bay. But violence was a shock to John Bahen. While he had seen union authority exercised and certainly been close to it in Oroville, California, the situation in James Bay was by far more serious and more dangerous.

Unfortunately, part of the infamy of the initial phase of construction of the James Bay Hydroelectric Project involves union-related troubles, including le saccage [rampage] de LG-2 where guns were brandished and expensive equipment was vandalized by angry union members. John describes one aspect: "Our equipment was parked at night, and we were not going to put guards out there to create a fight. So they just went at it at random. Out of $40 million worth of equipment, they damaged about $100,000."

Union problems touched Kiewit's roads jobs, due to two unions sharing dominance of the Quebec labour force: the Féderation de Travailleurs de Québec (FTQ) and the Confédération des syndicats nationaux (CSN). The absolute union control over hiring meant that no one could be hired, regardless of their role or skill set, without going through a union office. Rival unions sought a majority of work for their members on each job. Each union demanded overtime pay, which was the central issue in the disputes that provoked violent action on the part of some unionists. Contractors at James Bay were forced to reckon with the controlling and adversarial union bosses whose actions often included intimidation and violence that became personal. For example, John was shot at twice, and, for a time, when some union members tracked down his address in at 55 Foursome Crescent in Willowdale, he couldn't even come home. He phoned Margaret to tell her not to leave the house as there were two black cars ominously parked outside.

The events in 1974 eventually led to a judicial stop work order and Robert Bourassa's establishment of the Cliche Commission to conduct an inquiry into the problems of the Quebec construction industry and file a report. Its recommendations were later adopted by the premier and his government in early 1975.

The two men who had shot at John were found guilty and sent to jail. An ironic postscript happened many years later as John and Margaret were on vacation in the Turks and Caicos. Walking along the beach, they decided to stop at a bar. As the bartender turned around to take their order, there was instant recognition, followed by his very friendly, "Great to see you, John!" The warm greeting came from one of the shooters. Once released from jail, he and his fellow felon had bought the beach bar and were enjoying rather a nice life.

~

In May 1975, John's team was awarded its third major James Bay contract, Duncan Dykes. "It was an amazingly well-managed job and highly profitable," says Walter Scott, Jr. "While at the site with him one day, I said, 'John, we'll never, ever have another job as good as this one.' In the mid-1980s, when we were building Nipawin Dam for Saskatchewan Power, he reminded me of the comment and claimed I had been wrong. In dollars, he was right. But on a percentage basis, Duncan Dykes remains one of the most profitable projects in company history."

The contract for the project involved the construction of a series of sixteen dykes connected by over eighteen miles of gravel roads. These earth-fill, water-retaining embankments shaped the southern and western edges of the one-thousand-square-mile LG-2 reservoir, the largest reservoir in La Grande Complex. The head pond of the LG-2 Dam and Powerhouse, the largest in North America, holds six hundred ninety billion cubic feet of water.

Duncan Dykes remains the job that meant the most to John. He had a sense of arrival within Kiewit when the Eastern Canada District won the contract. He worked for a company that he was proud of and personally identified with, and, at Duncan, he was doing work he knew and was passionate about. He had the support of Omaha and believed that his team's efforts had firmly wedged Kiewit in the Canadian market by landing such significant work early in the James Bay development program. He had delivered on his promise in fewer than ten years. The profits were a positive for the company, and he was beginning to see the personal financial benefits of his diligence in finding appropriate work at the right price for his District. The Kiewit employee ownership model was beginning to build substantial stock holdings for him.

Bidding the Duncan Dykes required that the Eastern Canada District get some serious support from the engineering estimating function, senior management in Omaha and Peter Kiewit himself. At $138 million, the Duncan Dykes represented a financial risk five times greater than the Matagami Roads jobs had been.

Walter Scott, Jr., says John had an uncanny ability to draw people who possessed the skills he needed from elsewhere in the company. He then was able to convince their Kiewit managers to loan them for a job. Many unlikely individuals from the American divisions found themselves travelling to northern Quebec because John Bahen could "really use their help." For the Eastern Canada District, taking on work of this magnitude did not represent the slow and steady growth that Omaha might have preferred; rather, it was moving "Bahen-style," driven hard by its new district manager as each opportunity presented itself, regardless of size. He trusted that his team would rise to it and could count on the ample, trainable work force.

By the time Kiewit was bidding on the Duncan Dykes, Bechtel* was in place managing contracts for SEBJ. It had ramped up its input and was providing project management and engineering services. With the earlier roads jobs, however, SDBJ had hired its own staff directly for the specific purpose of contract administration. These "young rascals" failed miserably, in John's view. It was taking too long for payments to be processed. When Kiewit sought advice and decisions from SDBJ designers, answers were not readily forthcoming. When a change of conditions was submitted, SDBJ staff lacked experience to render decisions, risking profitability, something that John could not tolerate. It was this weak administrative environment, along with the complexity of building water-retention structures rather than roads, the size of the Duncan Dam project (physical and financial) and the uncertainty of the value that would be brought by Bechtel, that led Kiewit to look for methods to mitigate risk.

With the help of Omaha, John sought a joint venture partner to dilute some of these risks; that is, a second company, one that was not focused on being overly involved in the day-to-day, to invest financially in order to shoulder some of the burden of delivery of the work if things "went sideways." He found such a partner in Texas-based Brown & Root Ltd., with whom the Eastern Canada District pursued several contracts in James Bay, winning both Duncan Dykes and later the Eastmain Dam and Spillway projects. The joint venture, which worked out to 65 per cent Kiewit, 35 per cent Brown & Root, was known as Les Constructeurs KBR and established an excellent track record in James Bay.

John spent a lot of time in the area where the Duncan Dykes would

---

* The Bechtel Corporation acted as project management for the James Bay Hydroelectric Project from 1972 to 1985, which involved over four hundred contracts worth and almost $6 billion of work. Bechtel is one of the largest construction and engineering companies in the United States.

be built, walking the terrain along the centre lines of the future dykes. He carefully studied the geotechnical reports during the bidding period. He had read the bid document from cover to cover.

It was an estimate in which John's skill at delving into the minutiae – knowing the line items and specifications so intimately that he could find an opportunity for financial benefit – was a windfall for Kiewit. He was very talented at locating items where the tender document preparers might have underestimated quantities, incompletely analyzed potential outcomes or failed to remain open to significant change. While exploring the area, John saw that Duncan Dykes was essentially a massive earthmoving job. He describes how he took advantage of an opportunity: "When we got there, we dug test pits and we found the material we needed was closer to the dyke line. We changed our equipment scheme, so that our equipment went around in circles, and built the dyke in record time."

For the three hundred employees at Duncan Dykes, faced with a ruthless timeline and a short (May to November) work season, it meant double-shifting around the clock, six and sometimes seven days a week. John was still hands-on, much more than some of his direct reports might have liked. The back-and-forth banter over right and wrong, better and best ways to handle a particular problem was a constant on the job. John loved it, and, truth be told, his staff considered gaining ground on John with a useful argument or technical suggestion an amazing accomplishment, so they engaged.

John's favourite Pete Kiewit story is about the time that Kiewit visited James Bay in the spring of 1975, when Duncan Dykes was just mobilizing. It was "a disaster" in John's mind then, but, true to his positive attitude, he finds something to laugh about now: "We had left a pile of money on the table. In two days on the job Pete did not see anything like a productive operation. While investigating our single major risk on the job, the construction of a deep concrete cut-off wall which was still a year away, Pete got a 'one-pound' mosquito in his eye. He left early the second day. Alone!"

In June of the following year, Kiewit visited a second time and saw outstanding productivity, vigorous people, a safe job, good planning and below-budget operations. "He wasn't in such a hurry on day two, that time," John chuckles.

John admired the fact that Pete Kiewit, a wealthy American corporate leader, stayed in camp that night and met with men on site early in the morning in the mess hall. Later in the day, Kiewit donned a suit and met formally with Hydro-Québec people. When he was leaving Duncan Dykes, he was confident that John's vision of a very profitable job was in good hands, but Kiewit shared a story that gave John a rare glimpse of the closely guarded man he admired so very much. He told John about Guy F. Atkinson, owner of the company by the same name, who also lived in Florida and was a friend and golf buddy. They spent hours during golf games going over their companies' successes.

Atkinson bragged about the Mangla dam job in Pakistan, quoting something like $375 million as the contracted amount. He reminded Kiewit of the Atkinson coup and the corporate profit margin on the job. Kiewit grinned while telling John that Atkinson would go to Pakistan six weeks a year to check on the joint venture that made the company some $54 million. Kiewit admitted to John he had been envious at the time. But now, Kiewit exclaimed, walking across the tarmac to his plane, "Here we are two and half hours from Omaha making more money than Guy did in Pakistan – and Atkinson is dead now and I can't tell him about it!"

Pete Kiewit was an uncommon man. On one of his trips, when he was in Toronto with John, he developed a nosebleed that wouldn't stop and had to go to Sunnybrook Hospital. He was well cared for and, in a discussion with one of the nurses, found out that the Emergency Department needed some specific equipment. A week later, the hospital received a cheque to purchase the equipment.

John knew that James Bay was an exceptional opportunity both for the company and himself. Having Kiewit as a resource for seasoned people was key to his vision of maximizing participation in the James Bay program. The most critical piece in his long-term vision was recruitment, hiring, training and retaining young Canadian engineers to grow into Kiewit people and add to the construction industry.

Under John's guidance, his team members Bruce Daniels and Dave Callander co-ordinated an annual search for graduating engineers. Engineering schools were targeted, and, while those varied over the years, under John's watch the University of Toronto was always included. Other schools that produced significant talent to John's Kiewit teams included Memorial University, the University of New Brunswick, McGill University, Université de Sherbrooke, École Polytechnique de Montréal, Université Laval, Queen's University and the University of Waterloo. The recruitment program arguably started around the time of the Welland Canal jobs, but it gained significant traction in 1972 with the hiring of Donald Quane, and, in fact, continues to this day, although the list now includes schools from coast to coast.

Queen's University was a favourite of Daniels and Callendar in the mid-1970s, with some six graduating engineers joining Kiewit's James Bay team. One of these engineers was Ted Chant, an Anglophone Quebecer who had the same desire to build that is the root of John's own passion. The cover of the March 19, 1977, edition of the *Montreal Star* supplemental magazine, *Weekend Magazine*, showed a group of smiling men wearing hard hats, water-repellent overalls and filthy work boots. The caption read: "These men make as much as $1,200 a week working on the most expensive construction project in the world." The article, "The Power and Glory of James Bay," convinced Ted that what he wanted was both attainable and taking place on a mammoth scale north of his hometown of Montreal.

Kiewit had hired at Queen's the year prior to Ted's graduation, so he had ideas how he might target his intentions toward the company. Some research helped him conclude that it was a highly safety-conscious company, so he wrote his fourth-year thesis on "Safety in the Construction Industry" – something that, among a great number of applications, quickly caught the attention of Kiewit interviewers. Ted was short-listed by Kiewit, the only company that he sought an interview with. When John sat in on Ted's second interview, he went home and told Margaret, "I just met the man I would love to have as a son-in-law." Little did he know how prophetic that remark was.

Ted could not have had much of a grasp of Kiewit culture when he formally signed on after his interview and subsequent offer, in Toronto, in February 1977, nor that he would stay with Kiewit until 1996. He simply knew that he wanted to get building. This sense of urgency bode well for what Ted was about to experience upon completion of his studies on Friday, April 15. After writing his last exam and packing to return from Kingston to his home in Montreal, Ted thought he would take a month off before entering the work world, as his start date at Duncan Dykes was not fixed at that point. On the Saturday, Ted called the site to confirm details of his start. The woman who answered the phone was friendly and seemed receptive to his plan to start a month later. Two minutes after Ted hung up, however, John's project manager called from Duncan and suggested that if Ted was not on the plane north to LG-2 that Monday morning, he need not bother coming at all. This was Ted Chant's induction into the ways of Kiewit – and of John Bahen.

15

THE BURDEN OF JUST ONE ROCK

JOHN and his team shone in how they applied innovation. It was not acceptable under John's leadership that a job had to be built a certain way. This almost always meant questioning the wisdom of the engineers behind the design. John's approach leaned more toward "maybe we should try that," the opposite end of the continuum that starts with "we have always done it this way." He was willing to trust his team to bring forward the best ideas and to investigate his own ideas against the results of the district's past work; even more pivotal to taking on bigger and more complex work, however, was his ever-present willingness to seek help within the Kiewit corporate collective.

SEBJ had made some exact models of the Eastmain dam, showing how it could be built. According to John, "That was the biggest risk we were taking up in James Bay. The ground beneath the closure area was very soft, and the river flow was really strong as we were narrowing it down. It could have taken the whole bottom of the river out." So he called Martin Kelley, who ran the Omaha engineering and estimating team (and later became Peter Kiewit Sons Inc.'s chief engineer as well as a good friend of John's and Margaret's), to ask him to be involved in the model study work. In fact, he admits now that he made a "deal" with Kelley to give his undivided attention to the Eastmain contract when it was awarded. John saw the risk on the job and was wary of an earth-filled dam across a roaring river with an erodible bed. Kelley's presence provided the vast expertise that John needed in order to be comfortable with the bid.

SEBJ awarded the Eastmain contract to KBR in 1977. It represented another first – a dam and spillway structure. The Eastman-Opinaca diversions would augment the flow of the La Grande upstream of LG-2. *KIE-WAYS* reported on its progress in January 1978 in an article "Changing Course," saying that the job would be a joint venture valued at $58.6 million.

Although not as large as Duncan Dykes, KBR's Eastmain Project was a complex undertaking. The risks associated with each element, including closing a major river with a soft bed, was daunting. John took it all in stride. What concerned him most was building the spillway, because the construction of this type of structure was far outside his range of experience and professional comfort zone. He was also aware of the distraction faced by his team from the other risks that taking on the spillway job entailed.

John's response was to subcontract the concrete work. This was much to the chagrin of many individuals on the Eastmain team, including the project manager, Geoff Lindup. Until the Eastmain projects, Kiewit had not pursued structure work at James Bay. Lindup and others were aware that an enormous amount of structure work would come on stream in the years ahead, and they felt strongly that getting into the business of concrete structures with the Eastmain Spillway was opportune. John

remained adamantly opposed to concrete structures. A subcontractor was engaged.

Ironically, as it turned out, the subcontractor ran into serious performance difficulties late in 1977. The overall schedule, which required that the spillway structure be substantively completed to divert the Eastmain River, was in jeopardy. Reluctantly, John instructed the team to take over the structure work. This would be the first of many significant concrete structures that were eventually taken and delivered by the Eastern Canada District team at James Bay, which included LG-3 Intakes, LG-4 Powerhouse, LG-1 Spillway, LG-1 Powerhouse Phase II and the LA-1 Powerhouse.

When John and the team reviewed hydrographic information, they concluded that estimated river flows could make a July diversion possible (with some innovation), thereby putting the job back on, and possibly ahead of, schedule. It was the final cost that he was concerned with. For

The spillway bypass on the Eastmain project.

him there was satisfaction in a job that achieved its stated objectives of timing and price, but to improve on completion date and profit margin was what he lived for.

In truth, Eastmain was a high-risk dam to build. Once narrowed, the relatively slow-flowing river became a formidable threat to an earthen dam. A staged design was essential, so the dam design could be changed as the river increased speed and configuration. Geoff Lindup remembered learning a healthy respect for the dam's strength when he climbed down to the river's edge. As he describes it, "The water was so fast it was taking rocks off bigger than a trailer and just treating them like marbles, flicking them down the river. You underestimate the force, the velocity of the water. I wondered how we would ever get the thing closed."

The day arrived when the last rock was being put in place to divert the powerful river. Confidence was high by the time Walter Scott, Jr., visited the water's edge. John and Margaret had taken a helicopter flight north with Scott and his wife and the Lindups to watch as a fifty-ton truck dumped the last massive rock into the gap. While completions were great celebrations for all of the Kiewit work, the Eastmain closure marked the relief that the high-risk element of the dam build was ending.

John remembers looking at his boss's face, and, tongue-in-cheek, asking, "Why are you looking so concerned, Walter?" Trucks were crossing the built portion of the dam to dropping rock after rock to create a reservoir and force flow into the spillway channel. It was a remarkable moment, watching the flow of the river crash through the remaining gap and over the dam. As they waited for the last rock to stop the flow, Scott explained that, by his estimation, the amount of water could not be stopped by a single rock. "Even if the rock and the truck go in the gap, they won't be enough." But John had every confidence that more rocks could be added until there was an actual last one. Scott was simply pointing out the obvious, meaning that it would take as many rocks as necessary, until that final one stopped the flow. To lighten the mood, John said, "Walter, if the truck and the rock go in, there goes my son!" John's son Stuart was driving one of the trucks completing the rock plug that day.

~

After the Eastmain Spillway was successfully completed, John was feeling quite confident about his team's ability to take on structure work and, as a result, Kiewit, on its own, successfully low-bid the LG-3 Intake Structure. The value of the contract was $21.8 million, but, as with all James Bay projects, it needs to be understood what was included and what was not. SEBJ provided Kiewit with the concrete, the reinforcing steel, fuel, freight and camp accommodations (among other items), all at no cost. Adding only the supply of concrete and reinforcing steel into the contract value represented an additional $9.4 million.

Concreting of the LG-3 Intake took place over three seasons beginning in July 1978. Its successful completion set the stage for the next structure challenge to come, the even more daunting LG-4 Powerhouse.

John (left), Margaret, Carolyn and Walter Scott, and Geoff Lindup and his wife witnessing the last rock at the Eastmain diversion.

The award of the contract for the LG-4 Main Dam and Dykes in James Bay on November 7, 1978, was a milestone for the Eastern Canada District on a number of fronts, and a personal one for John as a builder. He holds up the estimating process for the LG-4 dam as a pivotal learning experience, and one with many instructive elements. Challenges learned from events going wrong or almost going wrong are equally valuable. In this case, one of the cardinal rules of competitive estimating had been violated, despite knowledge, expertise, good intentions and the all-important second estimate process. A total of $23 million was left on the table. In other words, the next closest bid by a competitor to their successful one was $23 million more. Money not made is almost as bad as money lost. There is an expression used in construction when big money such as this is left on the table: "Even if you are right, you are wrong."

While engineers are educated in standard estimate preparation processes, Kiewit training takes a basic education and refines the estimating practices to a much higher level of proficiency. Kiewit's unique second independent estimate process typically challenges three elements in the district or joint-venture bid: the right costs, the correct appreciation of risk and best methodology. What it gave John was "the confidence of knowing what the competition would be doing. Letting me see how many million difference there might be, and I would account for that difference in the margin we asked for. And the system almost always worked."

It was standard that Walter Scott and Martin Kelley, by then Kiewit's chief engineer, spent the last few days of a bid preparation period in Toronto going over details, especially when joint ventures were proposed. It reflected policy, not a lack of trust.

The bidding process for LG-4 yielded one of John's favourite stories about maximizing opportunity. When the estimating was underway, he committed his team to the two-estimate process yet again, involving Martin Kelley's team from Omaha. Donald Quane had been sent to

Omaha to work with the estimating team for two years, and he ended up participating in the pre-estimate document examination for LG-4 while there. The team of engineers had earlier projects to use for cost projections, there was an established inventory of machinery, and the list of experienced corporate staff members was in place. It was a straightforward process for the district team.

In his role in Omaha, Quane followed the same process as his colleagues back in Toronto, except that he did something that was not all that common: he read the French job specifications. He passed the knowledge he discovered in that review on to the Omaha team, as he had found a possible shortcut. At James Bay, the client SEBJ had established French-language priority when the project began. Contracts were signed in French, the language of the job site was French and, wherever the legal question of which version was to be followed in translation or interpretation – the French or the English – French prevailed. Quane found the text of the French version of the specifications allowed for building the dam in a shorter period than the four-and-a-half years stipulated. The subtlety of the wording difference in the English translation resulted in the English documents not allowing an early finish. The Omaha version of the bid was prepared with a projected completion of three years, whereas the Eastern District team's bid was calculated based on four and a half. It was the Omaha estimate that was used to bid for the LG-4 Dam.

LG-4 Dam and dykes was a $156 million project, making it the largest single dollar value contract undertaken by the district at James Bay to that point. The Omaha estimating group bid was a lot lower than the second bid. As an appropriate reward for his insight on the costs to be saved by completing more quickly, Quane (and his family) left Omaha after only a few months, rather than completing the minimum two-year commitment. Within hours of the job award in fall 1978, he was in James Bay in the role of project manager for the LG-4 Dam. As John told him, "If you know the job so well, you should get up there and run it."

Overseen by the over-achieving team and district, which had been accruing first-hand knowledge for six years in Quebec, with three major projects at James Bay, the LG-4 Dam was completed in three years. There were other innovations that paid in time and dollars. The well-established relationship with Caterpillar got the Kiewit job supplied with custom-built larger trucks than any other company would have taken the risk of buying or bringing to James Bay. Within Kiewit, if the machinery was not effective on site, it could be sold out to another district, minimizing the financial risk of custom ordering.

In addition, the project team developed an innovative, award-winning technique for cleaning dam foundations (a critical activity) with high-pressure pumps and vacuum trucks. The Canadian Construction Association awarded the Hugh R. Montgomery Memorial Medal to Les Enterprises Kiewit Ltée in 1982 for "excellence in innovation."

In construction lingo, John is a "dirt guy," not a "concrete guy." At LG-4 and after, the district began aggressively going after concrete jobs and became well known in the Canadian construction industry for the big concrete structures work. And structures work proved essential when the market slowed dramatically in the late 1980s and whenever smaller concrete jobs were the only ones around.

By 1980, ongoing bidding for work at James Bay yielded an award of the contract for the nine-unit LG-4 Powerhouse. At LG-4, some of the people John had been supporting, mentoring, challenging and elevating had their opportunity to rise to positions of even more significant responsibility. Donald Quane, who, as previously mentioned, came to James Bay straight from his graduation in Montreal to work on Matagami Road as a project engineer, was project manager for the Powerhouse job. His promotion took place by surprise one day when John needed a problem solved, and Quane left the dam project immediately. This was a common occurrence working for John Bahen. He says of giving his employees very little notice to relocate where they were needed, "Why wait? The building was going two shifts a day, six days a week."

The market in Canada had become sufficient for it to be run as a division on its own, with multiple district offices within it. Not long after John was promoted to president of Kiewit in Canada, Quane would be appointed district manager of the Eastern Canada District. Quane's career, spanning fewer than ten years from student to district manager for one of the largest contractor companies in North America, shows an impressive trajectory, one that almost rivals John's.

16

A FULL PLATE

IT might have been unthinkable to another leader, one not driven by optimism and a steadfast belief in his people, that the district could take on any more work concurrent with the massive contracts in James Bay. But, throughout the years of the heavy workload in James Bay, bidding continued, regardless of whether contracts were won or lost. John was not about to let work for the National Harbours Board (NHB) be turned away. He was always planning where people and machinery could go to be kept self-sustaining financially. Not long after the Matagami Roads contracts were let in the early '70s, Kiewit won the contract for Rodney Terminal wharf in Saint John, New Brunswick. John sent each of his best project managers to the jobs. John Loewen went to Matagami Road, and Geoff Lindup to New Brunswick.

The new cargo-handling facility was a forty-two-acre site for off-loading, storage and handling of loads from ocean-going ships and railroad cars transporting within Canada. The Saint John harbour is on the Bay of Fundy, site of the highest tides in the world. Between low and high tide, the water level rises up to twenty-eight feet. Marine construction in this context is extremely challenging, so, of course, it was just the type of project John enjoyed, even if he had very little marine experience. Work began in the spring of 1973 when a sub-contractor was dredging to a minimum of forty feet below low tide, preparing for the Kiewit work that involved driving octagonal piles to a depth up to two hundred feet of bedrock. A slope failure occurred in May, convincing first the contractor and then the National Harbours Board that the soils on the harbour bottom needed re-evaluation. Pile-driving halted so a fresh investigation of soils could be carried out. The slope failure had occurred on an edge where fill was freshly placed, releasing about five thousand tons of fill into the water. Lindup describes the problem this way: "There was a soft layer like toothpaste inside that had been restrained, and they didn't know about it, and, when they took the face off of it, out came the toothpaste!" Redesign caused a costly eight-month delay to construction of the wharves, a period of time when Kiewit machinery and a skeleton crew were basically sitting idle.

The redesign work was not the greatest problem for Kiewit, though, which had considerable experience with surprises in materials and time constraints. The greatest challenge of the Rodney Terminal job ultimately was cost recovery: for lost time, added work and changes to methods and materials, which were a departure from the bid documents. While Kiewit was not at fault, but rather was more at the mercy of the findings of the original subcontracting consultants and the dredging difficulties, it had incurred a substantial cost, and John was intent on recovering it. In his eyes, as with all jobs, this one needed to come in with a profit because Kiewit deserved it for quality work completed. His attitude is thus: "For me, it is basic math in business; you do the work on time and within budget, as contracted, and you get paid for it." To recover costs in this situation, however, meant convincing the National Harbours Board of this concept. It was John's favourite kind of argument.

The NHB was beyond budget and had no way of estimating what the eventual cost of Rodney would be. John argued that, unless the contracts were retendered, the documents in the bid package that was accepted by the client were valid. During construction, Kiewit kept billing at agreed-upon standard rates while equipment sat idle on-site. A major dispute ensued between the Eastern Canada District and the NHB when budget excesses grew as Kiewit invoices kept coming. The National Harbours

Board argued that it was under time pressure to complete the job for one of the national railways, and that the overage charges were unreasonable given the situation. Eventually, meetings in Ottawa arrived at a stalemate, which escalated to the point that John called Walter Scott, Jr., to come from Omaha to be involved in moving negotiations ahead.

"That whole job became a change order," says John. To him, it was irrelevant that the client, which was an agency of the federal government, found itself unable to balance its budgets or project manage efficiently. He argued that the fault should be placed squarely on the design engineers' shoulders, and that dependable contractors should not be punished in any case, when they had fulfilled their contractual arrangement. He further explains, "They didn't want to deal with facts because it would embarrass them."

Negotiations escalated, ultimately involving the deputy prime minister, at which time Scott made a deal for Kiewit that he could live with on behalf of Kiewit. John was never satisfied but respected that, at some point, every contention must be settled. Scott's take on the outcome acknowledges John's tenacity and focus. "We ended up getting some money, and it wasn't a great deal, and John always said that we were 'entitled' to it. There wasn't any question but that we were entitled to it. To me that is a great expression, which shows John's optimism."

~

In a joint venture with Northern Construction Company, the Eastern Canada District took the contract to construct eighty-eight miles of railroad between Gagnon, a company-created town, and Mont Wright, Quebec, for Quebec Cartier Mines. The contract was awarded in late 1972, and construction began the following year as soon as weather moderated.

The Eastern Canada District was stretched thin with Matagami Road underway. Participating in a joint venture gave John the comfort of diluting risk, but he did not want to lose control. The joint venture had been more the suggestion of the client than something Kiewit and

Northern wanted. They spoke about who had which expertise and people to lend to the project. John knew that Kiewit would likely have less to contribute than Northern from the standpoint of machinery and men. After all, they were building at James Bay at the same time.

Walter Scott, Jr., attended some of the initial discussions to support John in determining who the job's "sponsor" should be. The sponsor is the lead contractor; or put simply, the company assuming the greatest financial risk by percentage (but also netting the most financially). Northern could have become the sponsor of the job, as it had the experience and a lot of equipment available. But Scott was concerned with the percentage of involvement by Kiewit and went through the calculations carefully with John's team to ensure that it was even worth Kiewit's effort to be involved on the project at all. Scott's technique, presenting the facts and allowing the other company to draw its conclusions and expose its weaknesses and shortages first, was one that John admired.

Scott told Northern Construction that Kiewit wanted a decent percentage, but it did not need it to be too much, given that Kiewit might not be able to offer the same amount of men and machinery. Scott's goal, however, was to end up having more than half the percentage, even knowing that the client openly preferred Northern to Kiewit. It was a place to gain ground and profit without eliminating the joint venture partner, the presence of which was required by the client. It was important that the Kiewit stake just be more prominent, in Scott's eyes. In the end, Kiewit actually became the sponsor, and even Northern believed it was the best situation. Scott had argued well, and his show of patience as well as brilliant strategy was a style John admired and would later emulate at Hibernia.

These negotiating strategies became very useful not long after when Kiewit was low bidder on the LG-1 Spillway. Against the advice of some of his senior staff, John sought a joint venture with Quebec contractor Desourdy, which was regularly bidding against Kiewit on structures work. As he puts it, "I decided the best thing we could do to eliminate our competition was to make him our partner and also to teach him that he could make money on these jobs as long as you do everything right. If we got him with us he might see the light of day whereby he could bid these jobs at a reasonable price rather than being lowball all the time."

By 1975, John was deeply ensconced in Kiewit culture and felt that he had proven himself and was settled with a company that appreciated him and gave him ample opportunity to explore big building jobs. He had been promoted to district manager and was accepting more work than Omaha had ever expected. The first three James Bay contracts had been awarded, as had Rodney Terminal in New Brunswick. Matagami Road's miles were nearly complete, Duncan Dykes was Kiewit's to build, and, not long after, Eastmain River structures would be theirs. It was the district's busiest time since its inception.

That same year, the Royal Canadian Mounted Police (RCMP) made accusations and laid charges against John, personally, and Peter Kiewit Sons Co. Ltd., corporately. John faced conflict that would eliminate the tidy separation he kept between home and work. His name and address were regularly publicized, and the media aggressively associated him with what was then referred to as the largest and most expensive trial in Canadian history. He was one of fourteen individuals accused of participating in a Toronto-based bid-rigging scheme for dredging contracts on or near Lake Ontario. The four years that followed were very disruptive to his family and his career. Regardless of his not being found guilty, the trial made a permanent mark on him. The impact of the negative publicity of the charges and the strain on John are etched in the minds of Margaret and the children, who well remember the toll the pre-trial hearing and the formal trial took on him emotionally.

Accusations by the RCMP were based on the presumption that a co-operative relationship existed between the individuals and companies charged for the purpose of pre-arranging how each firm would set its price on a tender. The presumed price-fixing operation can be understood in this way: the contractors bidding on a given contract would agree on whose "turn" it was to receive work, and all other bids would be elevated at least ten per cent over the one whose "turn" it was, in order that they be eliminated. Contractors knew what government engineers had estimated for each contract because that information is supplied in the bid documentation, and they used that amount as a basis for estimates. The lowest bidder was always selected in the case of government contracts. Within the presumed bid-rigging group, the lowest bidder was responsible for paying some percentage or lump sum to the other companies whose bids were artificially inflated to guarantee disqualification.

The charges applied to sixteen projects; however, John and Peter Kiewit Sons Co. Ltd. were charged on just one count for the only contract that Kiewit bid on. John sought out the services of George Finlayson in Ottawa to act on his behalf before the Ontario Supreme Court. The relationship he had with his lawyer was ultimately frustrating for John, but it was only one of many frustrations. It is no surprise that the legal system was a source of irritation for him, given its process-heavy, time-consuming functionality.

John had many personal theories about whose statements might have led to his being included in the charges. Though it was difficult for him to stay composed, he consistently defended himself against the charges of collusion and never buckled under pressure. In the middle of the trial, Pete Kiewit flew in and assured John that he needn't worry, as Kiewit would pay all the costs. There was no question in his mind. He believed that John was innocent.

John testified that Kiewit's bid document was complete and submitted prior to the deadline. The prosecution alleged that John had received a call *prior* to his submission, soliciting his co-operation to bid high or not bid at all. Margaret, who had answered the phone, was even called in to testify on the key element of what time that phone call was received.

John felt that testimony on his charge heard in October 1978 significantly helped to exonerate him in the minds of jurors. That testimony referred to the instance when Kiewit had been low bidder on a fill job on Cherry Street on the Toronto lakeshore. Jack Jones, the Chief Engineer of the Toronto Harbour Commission, testified that he was glad that Kiewit was involved, because the Harbour Commission got the best fill price ever on the Great Lakes in that contract. However, the judge never referred to this in his summary.

Meanwhile, John's company was building mammoth and demanding contracts in Quebec and New Brunswick, and yet the typically very hands-on John was told by the judge, and reminded by him more than once, that he needed to attend court every day. Pete Kiewit told John that he needn't be in court the whole time. So he wasn't. "Fortunately," says John, "I was the guy rarely mentioned. The other guys all had higher profiles. And I wanted to keep it that way."

The status of John's alleged collusion in what came to be known as the "dredging scandal" or the "dredging issue" was unresolved for four years and, in the end, a "No Decision" verdict was returned on the charges against John Bahen. But some damage was done. John took the proverbial high road, realizing that he was totally supported by those in his company whose confidence he required and valued.

By the trial's end, eleven executives and nine companies were tried together, fewer than originally charged. John's former boss Harold McNamara was convicted, as were managers at Pitts Engineering Construction, Marine Industries and Sceptre Dredging.

17
BAHEN-STYLE

IN the spring of 1979, already district manager of eastern Canada for nine years, John was given additional responsibility as regional manager for both Canadian districts, including Les Enterprises Kiewit Ltée. The text of the announcement in the September-October issue of *KIE-WAYS* read as a short but impressive summary of the locations where he had participated in, or led, projects: "He has spent the past 30 years estimating and managing highway, marine, and heavy construction projects in Labrador, Manitoba, New Jersey, Ontario, New Brunswick, and Quebec. His responsibilities have ranged over equipment operations, estimating and bidding, and project and district management."

By the late '70s, John had an identifiable style. He had worked his way from the office to the job site and back, taking the best of Kiewit culture, policies and legacy. He was creative, detail-oriented, opportunistic and team-focused. He was challenged with growing the Canadian business and rewarded for his efforts in taking coveted contracts and completing them on time, on budget and, in every case, at a profit.

But these facts do not explain the complexity of John Bahen. He is an enigma, defying simple definition. His energy, outspokenness and ability to engage directly with people in a way that makes them comfortable are all part of that presence. His personality is also a reflection of his authority and confidence, and his belief that there is always a way to get something done. He is called tenacious and argumentative when defending his opinion, but he is also called brilliant, creative and inspiring. The "presence" of the man is often mentioned when someone is asked to describe his persona. Donald Quane summarizes it this way: "You know when John Bahen is in the room."

On the job site, two spoken or shouted words meant prepare for scrutiny: "Bahen's coming!" It was a respectful awareness that details would be examined and actions (read usually more work) would result. His arrival produced some trepidation in the staff he was visiting. A different reaction greeted the news of Peter Kiewit's arrival. When his helicopter landed on the job site, his visit was greeted with curiosity and a sense of awe. There are stories of people halting work, getting down out of the cabs of their trucks (which he put a stop to in his own humble way) and shaking the hand of the man whose name was on their paycheque every week – whose name was legendary on the job and famous in the industry.

Both as an engineer and as a person, John weighs situations and extrapolates possible outcomes; it is his way. Because he has given so much consideration to a problem that he can argue confidently and tirelessly for a given solution, he says, "You owe the guy you might be arguing against, having done your homework. He doesn't have to listen to you if you can't give him the facts and figures." In his mind, people need to see a confident leader, not one who is waiting for help to arrive, or who is willing to back down rather than champion the skill of his team. "Someone has to lead," adds Quane. "It is that simple, and John was always willing to take a position and stick with it."

Gene Bednarski, an engineer who worked with John for the better part of his career, says that they used to refer to him as often pursuing "pie in the sky" projects that no one could imagine – but many of which actually came to fruition and paid off handsomely, such as James Bay. "He was unique," says Gene, "and much more willing to take risks than most."

Gene has his share of "war stories" of working with John, particularly on bids. One was for Manitoba Power, which needed to divert waterways into a river system. To check out the site, John and Gene chartered a two-engine plane that had room only for them, the pilot and the man from Manitoba Hydro. Despite heavy fog, they flew north, finding their way by following the shoreline of Lake Winnipeg. Another plane was doing the same thing coming south, as Gene discovered when he glanced out the window to see it coming straight at them. The pilot "flipped" his wings just in time. Manitoba Hydro then arranged for a helicopter to take them closer to the site, where a plane couldn't land. Once dropped, they had to

traipse for a good three hours through bush and swamp. John pointed to planes buzzing overhead, saying, "There go the low bidders." They didn't know what was down there, but, because John did, and knew the challenges that could not be seen from a plane, his realistic bid was higher than the others, and he didn't win the Manitoba Hydro project. There was no point in winning the job if your bid was too low, because you would not make money on it.

Another bid took John and another Kiewit colleague, Gordon Smith, to Sept-Iles, Quebec, in a small helicopter to check out an eighty-eight-mile stretch of potential railroad. The pilot, checking his fuel gauge, announced it was time to go back. But John wasn't finished. So instead of cutting the surveying short, he had the pilot drop them on a mountain while he went for more fuel from one of the depots located strategically in this wilderness area. It was starting to snow, dusk was approaching and so were caribou. Luckily, the helicopter returned before things became too uncomfortable.

Even John's detractors, those who did not always agree with his pursuit of a particular process, admire his ambition, courage and relentlessness, because of his integrity. Says Walter Scott, Jr., "Behind John's trademark optimism and ready smile was an intense competitor who loved the contracting business. But what he loved most was his people. He was a demanding taskmaster and had what sometimes seemed impossibly high standards. But he also excelled at mentoring employees, in part by helping them to understand the talents they were not aware they possessed and challenging them to reach their potential. The combination made John an inspirational leader. His infectious enthusiasm and loyalty to his people were returned a hundred-fold. As a result, his team laid the groundwork for Kiewit today being the leading contractor in Canada. We're fortunate he found his way to Kiewit."

Jocelyne Desrochers, who has been John's assistant since 1977 and still works for the Bahens, admits that the stress level working for John could be high, but his staff felt that they did not ever want to let him down. And besides, everyone enjoyed being at work. He hired well, and he gave each of them challenges because he had confidence in their ability and simply gave them the opportunity, as Jocelyne points out, "to prove ourselves to ourselves."

Still, there was never a true downtime for "Mr. Bahen." Even when he was supposedly "relaxing"– travelling, or at the cottage – his mind was active and he was making lists and notes on correspondence that needed to be sent, the pile of which would land on Jocelyne's desk with a thud at 8 a.m. on Monday. She aptly describes his relentless pace this way: "This is the man who jumped down from the plane because he could not wait for the stairs to be wheeled up to the door."

His relationship with time was passionate and random. Time passed, and he continually tried to outrun it. When time was short, he simply went faster. When time was money, his feet did not need to touch the ground at all.

His preparations and tight timing as he left for business trips were legend. Using every available second, on a typical departure day, he would emerge from his office at 12 noon calling down the hall, "It's time to put the feedbags on, boys." This was the signal for a few of the engineers, perhaps Dave Callander, Gene Bednarski and Bruce Daniels, to join him, always at the same restaurant a block and a half away, for a working soup-and-sandwich lunch. At precisely 12:30 p.m., they returned to the office, at which point John made the mad dash out to the airport – for the 1:05 p.m. flight. The office was at Keele and Finch, some thirty minutes from the airport. Often, one of the employees went with him so they could work in the car, with John driving, then that person would drive John's car back to the office, or a second employee followed in his own car so John's passenger could go park his car for him unlocked, with the keys under the mat for John's return.

On another occasion, preparing for his next trip, he dictated correspondence to Jocelyne in the car on the way home, telling her to follow him through the house as he continued to dictate. This continued as he threw suits out of his closet onto the bed for Margaret to quickly pack, enlisting Jocelyn's help to close the suitcase by sitting on it. Meanwhile, he was standing at the front door, his hat on, calling to them: "Ladies, hurry up!" Dictation continued as he drove to the airport, with just minutes to spare, and dashed into the terminal, leaving the two women totally frazzled. Margaret then drove Jocelyne home, both of them too exhausted to speak. John, however, always remained calm.

Yet another common departure scenario was for John to pull up to the departures entrance of Terminal 1 on his own, jump out, grab his overstuffed suitcase and a clipboard from the back seat and, leaving the car idling and keys in the ignition, bolt into the airport. Later, he would call Margaret and ask her to find out where the car had been towed, whence she had to retrieve it and pay the hefty fine. His engineers also experienced the airport departure frenzy. Wanting to talk to one of them a bit more before catching a flight, John would insist that he or she join him on the fast and harrowing car trip to the airport and the jog straight to the departure gate (which you could do in those days). They would talk all the way, the engineer jotting everything down while running after John. Sometimes, when John felt they still needed to talk, he insisted that the completely unprepared engineer get on the plane with him.

Jocelyne also relates how they called themselves a "fly-by-night" operation – because staff literally had to fly out on the Sunday night to be at a job by Monday morning.

One of John's favourite pastimes was boating. Over time, his boats got bigger and faster; however, he did not really have time to learn proper navigation or the "rules of the road." As a result, he was known to ignore markers and become totally lost on Georgian Bay, or to travel down quiet channels too fast, creating wakes large enough to swamp private docks, interpreting the dock owners' angry gestures as friendly waves and simply waving back. Friends weren't always jumping at the chance to join him on his boats, aware that he had become known as "Two-Prop Bahen," typically going through two props a year. As John didn't have time to learn how to operate the radio, it was Margaret who had to take the test so that she knew how to identify the boat by name with the proper code. One day, travelling at his usual full speed across Georgian Bay, in his twenty-seven-foot Sea Ray, John failed to notice that the eight-foot inflatable dinghy that he was towing was swinging and dipping wildly behind, much like a kite. Oblivious to the reasons why people were frantically waving and pointing, he again simply smiled and waved back.

He may have become the "Big Boss" at many levels, but he still craved, as he puts it, being "one of the guys in the bunkhouse." For him, being a great builder meant still loving "what happens in the dirt" as much as he did as a boy. He held himself to the same policy and practice standards that others worked to. He was extremely loyal to the Kiewit way, and even more so to being John Bahen. He was not going to become an "office guy," which to him looked like "a corporate suit that sent orders out from a swanky office and never saw what was being built." He certainly did not want to become the visiting corporate English-speaker who came to Baie James and expected everyone there to speak his language. To his credit, he worked at being part of the job, which was what made him unique and what explains the respect he garnered from workers in James Bay.

The client at James Bay, SEBJ, operated in French. The predominantly Francophone workforce presented a challenge for the largely English-speaking management at Kiewit. But, in a move emulated by John himself, everyone attempted to use French as the job site first language. This was required by contract, but John enforced it as his own management standard as well, which undoubtedly gained him admiration from the Québécois workforce. A French-speaking job site was a first for Peter Kiewit Sons Inc. It became the norm for John, who, despite his intense and notable enthusiasm for the language, never quite grasped the correct pronunciation of some expressions, which became the stuff of well-meaning teasing.

As Walter Scott, Jr., says, "John's 'mastery' of French became a standing joke among employees. But while his occasional butchery of the language was humorous, it earned him a good deal of respect." He spoke their language rather than expecting them to speak his. One oft-told language-butchery story that John even retells himself is of his well-intentioned ordering of his favourite drink while in Quebec, a scotch and water – in French a *scotch avec de l'eau*. But his Anglicized pronunciation made the expression sound like *scotch avec l'oeuf*, which was scotch with an egg. It was entertaining for the Quebecers, but endearing more than anything.

Another story that people will remember about the work at James Bay was John`s commitment to having young people gain work and life experience in northern Quebec. He took the time to extend the tremendous opportunities of the Eastern Canada District work to the next generation. He believed then, as now, that young engineers need experience and, in particular, inspiration to maintain their belief that a career in heavy construction can be both profoundly impactful and successful. On the subject, Peter Kiewit said in a speech: "A company, or any other organization, will inevitably deteriorate unless it is constantly renewing from within – unless the right kind of young men and women are continually brought into the ranks, trained, directed and given every incentive to put their youth, vigor, initiative and resourcefulness into the business."

John found a way to mentor young people by bringing them north to be trained and work as productive members of the team on the James Bay jobs. It was not always easy, but he felt it was important to find a way to make it work while not jeopardizing others' jobs. He helped at least twenty young people obtain work through the summer that would pay for their return to university in the fall. These included the MacDonald and the Evans boys, Vere Mason's sons, the Irwins, Jeff Bertram, Alec Zimmerman, Peter White, Richard Carl, Jeff Hartley, Sandy Crawford, children of Kiewit colleagues, as well as John's own son Stuart.

Three fundamentals, which became part of John's personal lexicon and which friends and family can quote, having heard him repeat them so often, exemplified the Kiewit corporate guide to estimating and bidding.

The first is to "get work at the right price." For John to be completely pleased with the price at which his company obtains a contract, there should be no more than a dollar left on the table. Of course this never quite happened. The concept of money "left on the table" refers to the dollar value between one's winning lowest bid and the second lowest bid. If money is left on the table that implies that there was possible profit that Kiewit missed out on, and the client price would bear being higher if the second bidder is a substantial distance away in price. A one-dollar win over the second bidder is as good as it gets in John's world.

The second is to "build work at the lowest cost." This is one of the areas where John particularly excelled. The opportunity for profit need not end with the bid submission; John sought opportunities for cost savings on machinery, materials and "men," personnel within the work itself. This situation was vitally important in highly profitable work for the Eastern Canada District, such as Duncan Dykes. For example, he hired experienced supervisors who might suggest alternative methods that would produce additional cost savings during construction. He encouraged his staff to ask "what if" questions on the job, at lunchbox meetings or whenever a problem arose.

The third maxim – "take care of your assets" – is the most difficult to define. It involves both the mechanical, such as machinery, as well as the least controllable element – people. In this employee-owned construction company, the executives are engineers. They are builders, not accounting or business executives, or professionals from unrelated disciplines. Protection of this most important asset – its people – cannot be underestimated. This is demonstrated in the extraordinary attention to safety, the educational and training regimen, commitment to mentoring and internal promotion and the merit-based share purchase offerings.

Peter Kiewit fostered competition among the districts. He believed the presence of internal competition, in addition to market or external drivers, would unite Kiewit's people toward a common goal. He sought people, teams and districts that had limitless ambition as demonstrated by their enthusiasm. Forward movement was his catchphrase, and the reward structure was created to expand the talent pool.

Peter Kiewit Sons Inc. annual meetings are the vehicle by which its composite districts share knowledge of ongoing contracts and lessons learned with hundreds of invited employees. It is where presidents and executive team members deliver motivational speeches and announce forthcoming corporate changes. Long-term service and safety awards are presented, and group sessions on specialty topics such as building, equipment, excavation and structures are led by staff members. It is where engineers being mentored for promotion interact with managers. Sensitive topics such as motivating employees, avoiding litigation and creating safety consciousness are also addressed in sessions that are often led by staff members that one might not expect to be active at this level. It is at the Omaha Annual Meeting where corporate policy and corporate culture are most evident, and this is possibly the only time that Kiewit comes close to openly boasting of its endeavours, but only with the intention of inspiring its own people.

Held in January, the annual meeting provides districts with a summary of results for the previous year, compared to other districts. Each profit centre is responsible for presenting to the others. The announcements and awards are taken very seriously, because, in Kiewit, district financial results could mean personal financial benefit for a great many, in the form of stock offerings.

The Eastern Canada District's success within Kiewit was very evident when it received the Directors' Award for best construction and management performance for both 1977 and 1978. This was unfathomable for a fledging district, and much of it was credited to John and his anticipation and leadership in procuring the James Bay contracts.

In the 1977 session, Peter Kiewit's speech warned of the dangers of complacency and of the risks of looking back on accomplishments rather than forward to new opportunities. Given the dramatic growth and marked success of the Eastern Canada District since its formal creation in 1970, it might have been a time for John and the district to rest rather than buckle down and work harder. But hearing that message again at the annual meeting reinvigorated John, reminding him not to spend time dwelling on the district's excellence nor to risk feeling satisfied by the size of previous construction projects or financial successes. He came away from the meeting knowing that, for him, the greatest failure as a leader or as a builder would be to miss opportunities because of short-term arrogance. What made John so successful was his recognition of what it takes to build good teams: "A good leader sets examples, forecasts the future and motivates people by getting them to look at their achievements."

John receiving the Directors' Award, flanked by Walter Scott (left) and Bob Wilson.

18

INTO THE WEST

ONLY twelve years after being challenged to expand Kiewit's presence in Canada, John had offices across the country handling work in nearly every province and territory. Then, in 1980, he was appointed president, which was announced in the May-June edition of *KIE-WAYS* magazine: "The success of Peter Kiewit Sons Co. Ltd. has made it necessary to expand and reorganize the company. John Bahen, who was recently appointed president of Peter Kiewit Sons Co. Ltd., announced the establishment of district offices at Vancouver, Edmonton and Montreal."

This was evidence of the Kiewit Company's unbridled ambition that the Canadian arm of Kiewit would continue to dominate in megaproject construction in the '80s as it had in the '70s. Kiewit had been working in western Canada since the 1940s, but the work had been managed from a U.S. district while affiliated with Poole Engineering Company Limited, until 1955, when an office opened in Vancouver. In 1965, a Western Canada District was created with an Edmonton satellite office. Not long after, the estimating office where John began was opened in Toronto.

In March 1981, less than a year after being appointed president of Canadian operations, John was invited to sit on the elite twelve-person board of directors of Peter Kiewit Sons, Inc. and given the title of corporate vice president. The board had never included a Canadian appointee before (nor has it since). John was honoured and willingly assumed his role as an active board member, in addition to his duties and responsibilities as president of the Canadian affiliate.

John's assent to head Kiewit's Canadian operation initiated a change in how Kiewit pursued and managed work within Canada. While eastern Canada, under John's leadership, was successful in a grand sort of way, western Canada, under the control of John Patterson in Vancouver (before John Bahen's arrival), had a record of solid but not spectacular success. In Edmonton, under Ernie Elko's leadership, Kiewit was focused on road construction and paving, the returns for which were gaining steadily. While not experiencing a loss, these enterprises were not remarkable. Safety was excellent, however, and the operations in western Canada had a strong reputation for doing things the classic Kiewit way.

When John took over nationally, he offered the two western Canadian teams an opportunity to achieve heights not possible before. Edmonton landed the Dickson Dam project in 1980, a significant flood control water-retention structure on the Red Deer River in Alberta.

During the later 1980s, with John's encouragement, Vancouver flourished under the successive district leadership of Lee Kearney and Rod Bulloch. A significant job that the Vancouver District took, sixty miles west of Lethbridge in southern Alberta, was the Forty Mile Reservoir Project. This is one of the driest areas in Canada, with low annual precipitation. Awarded in 1985, at $36 million, it was the largest of several projects undertaken by Kiewit for the St. Mary River Irrigation District during the mid to late 1980s.

Like Dickson Dam, the Forty Mile Project was being managed by Alberta Environment. In both cases, John believed that Kiewit was entitled to additional compensation for work done in delivering the projects. John championed both causes, although at Forty Mile he had a passionate partner in Rod Bulloch. Both claim processes proved Kiewit to be right in part. John was not satisfied with either result, stating on many occasions that he was pleased, but not satisfied, with the outcome.

John was personally engaged in both projects for many reasons. The simplest is that both were his preferred type of work – earthmoving. But he had other motivations that display his style and character as a leader. John was instrumental in the strategy that inspired both districts to reach beyond their comfort levels to get jobs outside their typical work, and he wanted to continue to support the approach by staying close. In both cases, the project managers were young eastern Canadians, Ted Chant and Mike Devine, whom John had hired himself and in whose careers he had played a pivotal role by assigning them to major work. As was his style, John wanted these two individuals to succeed. At Bow Island, Alberta, the closest major urban centre to the Forty Mile coulee site, he found perhaps the most powerful reason to be personally involved in the western Canada projects. His daughter, Susan, by then married to Ted Chant, and their daughter Katie, lived at Bow Island, Ted having been assigned to Forty Mile as project manager in December 1985.

In the early 1980s, young engineering hires by John for the James Bay jobs yielded a large contingent of successful lifetime Kiewit employees, many of whom rose to senior positions in the company today. One of the most successful jobs in Kiewit history was the Nipawin Dam project in Saskatchewan (now called the François-Finlay Dam and Power Station). The impetus behind the 252-megawatt power plant was to provide an alternative energy source to fossil fuel for the province of Saskatchewan. For ten years, it held the title of achieving the highest gross dollars in profit for Kiewit.

Some thirty years later, John still takes pride in many aspects of the Nipawin job, and the financial success was only part of it. He was proud of the relationships built between Kiewit and the community via co-ordination with municipal government and the mayor's office in the town of Nipawin. Leaving a positive lasting impression of Kiewit in the community where it builds for a few years is a personal hallmark of his. He was able to directly transfer many Quebecers, and other French speakers, from James Bay to Saskatchewan where there was a small French population, and, in that way, to provide a sense of their roots for many workers.

It was at Nipawin that many young engineers moved up into positions of greater responsibility. Nipawin, however, was a megaproject, and, by the time it was completed, in 1984, the market was not bearing that type of work in Canada. The team of managers and supervisors was dispersed at that point, but hires under John's leadership still represent the majority of upper management in the Eastern Canada District.

In the lore of John Bahen, there are innumerable stories of his grasp of complex situations and his ability to render simplified solutions, and there are as many stories of how he demanded the same from his team. John recognized the importance of the Nipawin project to his operation at an early stage, which was only one of the reasons that he insisted that his "A-Team" from the Toronto regional office visit the site in advance of the Omaha estimating team.

Prior to estimate review, Dave Callander, Gene Bednarski and Ted Chant joined John at Nipawin, a trip that involved a three-hour flight from Toronto and a three-hour drive from Saskatoon. The three engineer-estimators studied the plans and specifications well in advance of the trip. They had varied experience and seniority among them, but all felt confident that the issues presented by the Nipawin project were recognized fully, including issues on the progress of the drainage tunnel, the predicted construction industry strike, competing bidders and challenging key physical features. John's well-known rigorous style was to subject his staff to the same probing questions that they could expect at the review stage. He needed to know that his team would anticipate a problem and include a solution within the bidding documents. John knew that they needed not only to work harder than everyone else in order to get the Nipawin project but to work smarter as well.

Ted Chant recalls the group standing on the left bank of the Saskatchewan River, at about the point that would become the centre line of the required by-pass channel excavation. With about forty-eight man hours under their belts of reviewing the drawings not to mention about seventy-five years' total experience among them, the three engineer-estimators nevertheless had to answer a question that had occurred to John while he was standing on the riverbank. They knew that the purpose of the excavation was to create the necessary materials for the Main Dam embankment. John glanced across the river and said, "Guys, if the material for the dam is on this side, and we are diverting the river through the middle of the site, how do our trucks get to the other side to build the dam?" Callander, Bednarski and Chant had no response. Walking back to his vehicle, John turned and stated, "I guess we'll need a bridge. You need to put that in the estimate."

A joint venture bid by Kiewit, Guy F. Atkinson Construction Ltd., Commonwealth Construction Company Ltd. and Ramsey Construction Company (known as K.A.C.R.) successfully bid for the $208 million main civil contract to build the Nipawin Dam for Saskatchewan Power. The contract was awarded in the spring of 1982, within days of a pivotal provincial election, which turned the tide of politics in Saskatchewan. On April 26, 1982, in the Saskatchewan general election, the Progressive Conservative party defeated the New Democratic Party, which had held power for over a decade. Grant Devine's majority victory was historic in the province. It took months of demanding negotiations for John and his management team to convince the new premier that he should honour the commitment of his predecessor and continue with Nipawin (and with K.A.C.R.), before mobilization began. Upon reflection, though, he simplifies what this entailed.

"The government changed," he says. "The contract was already ours. There was a delay, but they didn't try to take it. We were the low bidder, but they weren't used to acting quickly in Regina. They just wanted to think about it for a week or two."

Not only did John and his people need to overcome the politics of the day, but, even with the submitted low bid, the project cost was greater than what SaskPower had anticipated. SaskPower's Bob Lawrence led the negotiations on the owner's side, and John led for the joint venture, with assistance from Kiewit's nominated project manager, Gary Tackett. It was an "on-again, off-again" process that stretched out longer than usual. Few of those closest to John believed that he would ever give up on it, so they were not surprised by the announcement that the Saskatchewan legislature called the project a "go" and acknowledged that K.A.C.R. had been awarded the contract.

Receiving the contract for Nipawin was good timing for the Eastern Canada District. John had a lot of people and owned equipment coming off the LG-4 Main Dam project, and moving from northern Quebec to northern Saskatchewan was reasonably efficient for both. He also knew, and was hearing from his competitors, that the flow of public sector work was slowing. The size, timing and scale of Nipawin made it a job that John knew his district needed to extend its level of activity. Nipawin would ultimately last almost four years, carrying the district's staff through a large part of the impact of the recession happening at that time. Ted Chant replaced Gary Tackett (who returned to Omaha) at the end of the third year of the project's construction period.

There was another circumstance that would ultimately contribute to the success of the project. A unionized construction strike in the summer of 1982 not only delayed the tunnel work that was required prior to the start of any dam construction but also threatened to delay K.A.C.R., because the job was tendered as a union undertaking. John and Gary Tackett made the decision to build the work non-union. This would not have been possible under the previous provincial government, and John sought assurances from Premier Grant Devine that K.A.C.R. would not be penalized for adopting an "open shop" strategy. John received positive, if confidential, assurances and work proceeded. It was not the wages or working conditions that differentiated union and non-union workers; it was productivity. A number of factors that made union-built contracts less productive included delays caused by arguments over jurisdiction between competing unions, "feather bedding" (where unions dictate the number of workers required to complete a task) and the fact that union members, unlike open shop workers, could complete only a single trade task. Perhaps most important, at the time of the award of Nipawin, Saskatchewan was in a deep recession. Much of the skilled and younger labour had left to work in Alberta on the tar sands. Without the hiring hall restrictions, locals returned to Nipawin to work on the project.

The work ethic in rural Saskatchewan amazed John. Many of the productivities delivered by local people in Nipawin, especially given that most of them had not done this kind of work before, were some of the best that Kiewit had ever seen. A testament to the importance that employing local people played in the overall success was the underutilization of the proposed eight-hundred-person workers' camp that ultimately installed only five hundred beds. In fact, the camp never exceeded a population of four hundred, a shortfall directly related to the participation levels of local workers. Crews typically worked two ten-hour shifts a day, every day but Sunday.

Although it was evident throughout his career, John's passion for equipment and building was never more obvious than at Nipawin. Perhaps his interest in equipment explains his fascination and focus on earthmoving work throughout his career. He loved heavy construction equipment. Seeing equipment being productive was the best feeling of all for him. He firmly believed that the right answer to any equipment selection question was simply to use the biggest of the proven equipment that suits the job. While John was criticized for a single-minded drive that often meant overriding the opinions and input of others, there was one thing he was admired for unequivocally. He firmly believed that equipment must be well-maintained and that it must always appear to be so well-maintained.

In fact, equipment that was transferred out of John's operations to other districts had a reputation of being exceptional in maintenance and appearance. Pickup trucks never left the district without new tires and spotless detailing. Surplus equipment was repaired and repainted. The Forty Mile coulee Caterpillar scrapers went on to the Northern California District. A letter from the district manager in California to John's district manager in Vancouver stated that never in his long career had he received equipment that looked as good as what came out of the Forty Mile job. John would not have his reputation reflect anything less.

John's knowledge and attention to detail were evident even in his less formal work dealings. He was always "on." One night in the bunkhouse at Nipawin, over a card game, he shared a story that he could not have known would conclude with his appearing in court for one of his competitors. John was famous for his stories, most of them about excavation, cubic yards being moved and equipment change outs, and his listeners usually benefitted from them. This time, John told the card players, who worked for Atkinson, that Kiewit was doing some work in Fort McMurray, Alberta, for Syncrude. He said how hard the work was there and complained about the ground and having the wrong machines on site owing to the radical changes in soil conditions from what he had been led to believe by Syncrude. He told them that the job was a financial roulette game, but he was stuck with it because it was a ten-year commitment. When doing another Syncrude-led site tour with prospective contractors a short time before, for other work, he found that Syncrude engineers guided the visitors only to the sandy loam area and not to the area where the heavy and abrasive clays were common. He thought Syncrude was being deliberately silent on the exact nature of the geological conditions.

John said he saw that his Kiewit employees were having more difficulty in excavating than expected, largely because the wrong equipment had been allocated for the job. The operation covered many miles, with pits being quite a number of miles apart. Geologic conditions changed over those distances from sandy loam, which is what was expected, to a very wet clay, "which was like a bear cat to move" for Kiewit's equipment system. The heavy clay cost three times more per cubic yard to move than the loam, which the contract was paying for. The terms of the contract were radically different than site conditions, he said, and Syncrude knew it.

"Things in the game are starting to look up, and all of a sudden one of my partners from Guy F. Atkinson Company mentions how tough things are for them working for Syncrude. And I said, 'Well, haven't you guys discovered that Syncrude are pulling the wool over your eyes?' They said no, and they wanted to know. So I told them about our experience."

They then asked John to talk to their management, and encourage them to confront Syncrude, because they were sure that the wrong equipment would be moved onto their job, too. A year later, John was served a warrant to appear in court on behalf of Atkinson. He describes it thus: "So I appeared, much to my concern about my and Kiewit's future opportunities, and, sure enough, about ten minutes after I told my story, the case was dismissed. And Atkinson was awarded some money."

~

The decade of the 1980s was an eventful time for John's family. He was more mobile, yet his desire for balance of home and work was continually challenged in the early part of those years. He and Margaret had purchased a home on Abaco in the Bahamas, which made short-term getaways possible for everyone, thus preserving some family time. This happened because Susan had gone there with friends after her graduation from Guelph and loved it. The Bahens rented a villa the next year, then bought a house. By then, Susan was studying at Hofstra University in New York.

For Ted Chant, there was an unusual twist to the work life/personal life division. At that time, he was still working at James Bay and was about to become the Big Boss's son-in-law. He had met Susan when she went with Stuart to a football party at Ted's place. When he subsequently came

TOP OF THE ROCK
Treasure Cay, Abaco, Bahamas

*John teaching Margaret's father, Alex Campbell, how to sail in Abaco.
House in Abaco, Bahamas.*

to the Bahen house to pick her up for a movie date, John said, "What are you doing here?" Ted simply responded, "I'm taking your daughter out." Despite the daunting prospect of dating (and later marrying) the boss's daughter, he went ahead.

There is a classic story of Susan announcing her engagement at Abaco in 1982. She recalls that it never was exactly easy to get her father's mind away from work or his Kiewit green leather pocket calendar. No one was surprised at her news and all were overjoyed. The natural second step after the congratulations was to set the date. Susan and Margaret went to John, who immediately checked his little calendar. He concluded that he did not have a weekend free until the next Christmas. Susan and Margaret tolerated his thought process for a few minutes, expecting that he would realize that his needs could not possibly outrank a mother's and daughter's wedding planning. They announced what they felt was a reasonable date in October and let him know that if he found he was not able to make it, they were going ahead anyway.

The entire family was there in October of that year. Stuart came home from his work in California (for Kiewit), where he had gone after graduation in engineering from Queen's University in 1980. Michael had completed his schooling at the Phelps Academy and was living in Toronto.

Michael had epilepsy. He also had Attention Deficit Disorder (ADD), which wasn't diagnosed in those days. As a result, his personality and behaviour could be challenging. Margaret knew, when Michael was very young, that something was not right. It was difficult to explain to John a condition that had no clear diagnosis, but there was little they could do – except to deal with it. Michael was a very determined young man. While at Phelps, despite having been told by his parents never to get a tattoo, he and three other students went into Philadelphia and did just that, his consisting of a skull and crossbones and a Nazi symbol. The boys were sent home for two weeks. Margaret, who already knew why Michael was coming home, advised him to wear a long-sleeved shirt. He didn't. When John picked Michael up at the airport and saw the evidence, he gave Michael three choices: he could cut off his arm, leave home or get rid of the tattoo. Michael opted for the third choice. He was not aware, at the time, of the meaning of the Nazi symbol, and his parents were concerned that it could get him into trouble as he got older. Consequently, John and Margaret spent time educating him about the war and Nazi symbols. Regardless of the challenges, John was a great father to him.

In August 1984, the Bahen family suffered the tragic loss of Michael, his death caused by a grand mal seizure (known as Sudden Unexplained Death in Epilepsy, or SUDEP). He had been working at Goodwill and living on his own. His family was unaware that some epileptic seizures could be fatal. Susan came home from Saskatchewan and Stuart from California. They stuck very close together and spent some time at the cottage before the children left. John was inconsolable. Margaret, supporting everyone through the agonizing grief with her inner strength, suggested to John that he should return to work. Having volunteered at North York General Hospital, where she successfully ran the gift shop, Margaret and four of the women who had worked with her had then opened a gift shop in Unionville. She left that business and travelled everywhere with John. That is how they managed to move forward together.

Since then, Margaret has devoted much time and energy to the Michael Bahen Chair in Epilepsy Research at U of T, helping to choose both Chair holders. More recently, in addition to the Bahens' past support, Margaret and her daughter, Susan, worked with the Campaign Cabinet to raise substantial funds to build the Cancer Centre at Southlake Hospital in Newmarket, for those living in York Region.

*The Bahen family on the day of Susan's wedding.
Standing, Michael (left), Susan, John. Seated: Margaret and Stuart.
(Photo: Brant Wilson)*

19

PLANNING THE NEXT MOVE

THE economic recession began to show itself in the Canadian economy and the construction sector by 1981. In Omaha, Kiewit was forecasting a change in the construction market as early as 1977, when Bob Wilson wrote in *KIE-WAYS* magazine that an economic change was no reason to depart from the basic tenets of Kiewit's competitive strategy. "We have seen these conditions before, and, as in the past, the work will go to those companies that are the most efficient and that operate at the lowest cost." Aggressively dispelling fear in his message, Wilson reiterated the policies that had enabled the company to survive and thrive. "This is not a time to expand our commitments on work that is bid unrealistically for the sake of volume." In addition, he warned managers not to lose able, trained people because the work volume was reduced, nor to pursue types of work with which they had little experience because of the slow period. "Rather, it is a time to pursue our estimating and bidding effort on a sound, profit-oriented basis." In a speech the following year, Peter Kiewit said: "The progress of our company in the past has been made possible by our ability to adapt ourselves to new and changing conditions."

John observed the effect that a varying amount of public sector work coming on stream across Canada was having on Kiewit business. Donald Quane was wary of what the recession might mean for Quebec and for the work that had been flowing from Hydro-Québec. It is managers who must concern themselves with finding new work and planning a few years ahead. "We worry first, hoping no one else will have to," says John. Both men were aware that ten years of nearly limitless public sector work had to end. The incredible influx of work in James Bay had demanded every possible resource, and there had not been time or energy to procure jobs in the private sector. The steady flow of massive projects was such a focus of attention that, with Phase 1 being projected at a fifteen-year build-out and Phase 2 following on its heels, there had not seemed a need to prepare an aggressive "Plan B."

There was little major work in Quebec and Ontario, so Kiewit took small contracts to make use of people including sewers, roads and some water projects. They excavated, built small structures and did some earthworks. But the work between 1982 and 1989, while Donald Quane was District Manager, was quite small compared with the James Bay work.

This was a time of belt-tightening and, for other companies facing cutbacks, there would have been extremely low morale. At Kiewit, the impact on morale may have been minimized somewhat because almost all of the supervisory staff of the Eastern Canada District at that point were stockholders. Stockholders are more tolerant of the "feast to famine" periods that construction is faced with both seasonally and periodically.

People who own stock understand that a profitable income and a project that completes on time will give their district and their job the longevity they hope for. John remembers this time not as a period of recession but of working at a normal pace. This might be explained as classic Bahen optimism, but also as reflecting his intimate knowledge of both the market and the construction industry. He knew that the size of the projects and the amount of the fifteen-year budget of James Bay would never come again, and he had capitalized on it. When the rebalancing of the 1980s came, he looked at it as a period in which any cockiness they might have developed would be purged, and any laziness that resulted from the seeming ease of getting massive contracts would be cured.

John was very good at seeking people with particular skills and experience from elsewhere in the company. When there was less work in the '80s and attention was focused on employing people and equipment, transfers for months at a time within corporate Kiewit became the norm. When other district offices, especially within Canada, sought some help on a particular job, John authorized the transfer between districts either in Canada (his preference) or to the American districts. The transfers were established as secondments only, as, after ten years in James Bay, his team of supervisors was very valuable to him. John was anticipating future work there, so he wanted to be sure his staff came back.

Susan remembers a postcard her father sent from Prince Edward Island when he travelled there for business. She was enchanted by the notion that her father was in the fictional home depicted in *Anne of Green Gables*. He wrote to her that it was very sunny, and then (ever the engineer), noted casually that the place could "really use a bridge." The three-hour ferry ride from New Brunswick seemed ridiculous to him in terms of both time and cost.

John was not the only person formulating ideas for how a fixed link between Prince Edward Island and the mainland could be constructed, one that could resist currents and weather and traverse the huge distance of the Northumberland Strait, which is icy more months of the year than it is free-flowing. He was contacted by a lobbyist in New Brunswick representing interests that desired the bridge, which fuelled his fire to be the company behind the eventual bridge. Former employee Colin West says John asked him to draft a letter that would be sent to politicians, local councils and citizens' groups in New Brunswick. Then, for the initial bridge proposal to Ottawa (not the bid itself, which was years later), John asked one of his employees, Mike Devine, to go to Ottawa to research the Parliamentary minute books regarding the wording of the government's proposed responsibility to PEI of maintaining a permanent link (ferry) to the mainland once PEI joined Confederation in 1873. It was all about the details. He directed staff to prepare a proposal for a steel-based bridge design. And it had to be ready the next day. This was not the first time they had been directed to prepare a comprehensive document to a tight deadline. What this meant for many of the staff was that they would be working through the night and that when it came time to collate the document, which was typically several hundred pages, everyone would gather in the boardroom and, regardless of their position in the company, share the task of assembling the books. John presented the proposal to politicians in Ottawa, met with people in both provinces and ultimately provided the staff to educate the public before a public plebiscite.

Colin remembers the day that he was listening to CBC Radio while driving and heard the familiar voice of John Bahen passionately discussing how a bridge connection between the mainland and Prince Edward Island could improve not only transportation connections and job opportunities but also the overall quality of life for island residents.

Ultimately, Kiewit was eliminated at the pre-qualification stage, along with some of the largest contractors, including Bechtel, which had administrated the James Bay Hydroelectric Project. While the eight-mile curved Confederation Bridge is now applauded as an engineering wonder, John had wanted steel while other proponents expounded on the engineering virtues of concrete.

Nevertheless, John Crosbie, a federal cabinet minister from Newfoundland, had applauded John and his efforts and said the bridge should be built by Canadians.

~

Peter Kiewit Sons Inc. proudly announced a golden anniversary of construction in Canada, in the September-October 1991 issue of *KIE-WAYS*. In a feature by president and chairman Walter Scott, Jr., "The Maple Leaf Turns Gold," Scott describes John's key role in the district that has "constructed some of our largest and most challenging projects in the past two decades." He later said, "John probably was about as

KIE-WAYS

SEPTEMBER-OCTOBER 1991

LG-1 Spillway in Northern Québec

**KIEWIT
50 Years of Building in Canada
1941-1991**

persistent a character as I've ever run into. He would chase a detail down, you know, regardless of how long it took to get it, particularly if it had a bearing on how the work was bid or how the work was being built. He was extremely detailed and persistent about making sure the details were covered."

No matter how long John was involved in the construction industry, the reminders of basic lessons were relentless. During the Hibernia project, John relearned the need for a company to possess design expertise and innovative construction methods in order to be indispensable. The inability of one contractor to grasp both elements was ultimately why Kiewit got involved in 1994 in a joint venture with Aker Solutions, a world leader in concrete platform construction. This project, which John comically refers to as a "project rescue mission," is what took him to Newfoundland and Labrador.

Kiewit's involvement with the Hibernia project began with pre-bid discussions. The largest construction project in North America in the 1990s, Hibernia is an oil drilling, production and storage facility colossus. It is located 196 miles southeast of St. John's, in the area of the Grand Banks, fixed to the ocean floor. Hibernia produces one hundred fifty thousand barrels of oil a day. The $5.3 billion megaproject (or perhaps best called a gigaproject) had a concrete caisson job at its core. It was the scale of project that John could not ignore.

He took a team of people down to Texas to present the "dog and pony show" and was bluntly told by representatives of the Hibernia Management and Development Company Ltd. (HMDC), the consortium that owned Hibernia, not to bid on the contracts for the gravity base structure (GBS) of the off-shore rig. John sought clarification as to whether Kiewit had failed to pre-qualify, perhaps with echoes of the failed Confederation Bridge bid in his head. He was furious at being told that Kiewit would be wasting its time to bid. He learned that they were not formally disqualified but rather had been ruled out because Kiewit had never built such a structure in the North Sea. Unwilling to be backed off the largest construction project of the decade so easily, he made numerous attempts to argue both the Kiewit Canadian capabilities and to guarantee that the resources of the entire corporation would be made available as needed. Still, the HMDC project leads refused to receive an estimate from Kiewit.

He smiles now to think how upset he was, but he had other reasons to grin. As it turns out, he had the last laugh. A year and a half after the contract for the Gravity Base Structure had been awarded, Kiewit was asked for help in moving the project forward. The main owner acknowledged that the European consortium was not able to design and build the GBS. Cost overruns were only one component of the owner's concern; the other was the base construction coming in late.

Another rewarding aspect of the Hibernia project for John was that the owner had previously selected a site that John felt was less than ideal with respect to facilitating construction of the project. On a wickedly cold and wet Newfoundland Sunday, Donald Quane and Dave Callander were in a small fishing boat that John had chartered with the express purpose of finding a better dry-dock location for fabricating the GBS and assembling the topsides, then mating these two large components, after which the structure could be floated and towed to its deep water home. John chose Bull Arm for the construction, and, a few years later, after his retirement, the structure – all four hundred and fifty thousand tonnes of it – was in place in the Atlantic.

20

TIME TO GIVE BACK

JOHN BAHEN joined the Kiewit board of directors during a time when his career was on a rapid ascent. His aspirations were well served by his role as president of Canadian corporate operations and a seat as the first – and only – Canadian on the exclusive Kiewit Board. Thirteen years later, by 1994, John was in a different state of mind. He was still passionate about the profession, but knew that there were years ahead of him when flexibility in schedule and more personal time would be healthier, however unaccustomed he was to both. He still had projects that he was dreaming of, boats to look at, grandchildren to play with, a golf game to master and the world to travel with Margaret. He began to reduce his commitments to strategic and long-term planning with the board, ultimately resigning his duties.

While others might have taken years to orchestrate their departure, John did it in calculated steps over about a year, starting with retiring from the board. Some people work to live, until a time when work is less important financially and personally. John would no more want to be content than to be described as "retiring"! The fact was, at Kiewit, retirement age was sixty-five, and he knew there would be no exceptions. In order to have a couple of irons in the fire, he worked with business colleague Joey Tanenbaum as an investor and member of the board of Brascan to develop and enhance hydro developments near Ottawa for Hydro-Pontiac, of which Stuart was General Manager.

When it came time to retire, his greatest challenge was going from

John and Denis Evans golfing in Jasper.
John and Margaret in Chile.

In Sicily and Egypt.
John's plane, with call letters ending in JB.

a near manic pace to unplanned time. John's reaction was to engage in a meticulous travel regimen, consisting of research, itinerary planning and the development of serious travel milestones to be achieved, including golfing the best courses and seeing the yards of the finest boat manufacturers in Europe. The Bahen kitchen became travel planning central. Whether it was visiting boat builders in Germany and Holland, enjoying the culture of Italy and the pyramids of Egypt or savouring a cruise from

In Kenya.

Prague to Budapest, all told, they travelled to forty-eight countries in John's first few retirement years.

They were flying here, there and everywhere so much that they finally decided it was more practical to have their own plane and, in 1998, bought an eight-passenger Beechcraft King Air 350. Besides the relatively short-haul trips in North America and to the Bahamas, they embarked on the trip of a lifetime to South America in 2000. This four-and-a-half-week trip covered over sixteen thousand miles through Ecuador, Chile, Argentina and Brazil. As chronicled by their pilot, Don Merchberger, in the industry publication *UniCom*, they flew past an active volcano and missed a presidential overthrow in Ecuador, met penguins in the Antarctic, landed on isolated air strips, hiked around Machu Picchu and "visited some of the most magnificent and exotic wonders of the world."

~

Due to John's hard work and Kiewit's wise investing, the Bahens not only embarked on some exciting trips but were also in a position to commit to some major philanthropy. They focused on two main areas: health care and education. These two themes keenly reflect each of their philosophical passions, and they continue their support to this day.

Health care had long been Margaret's forte. Given her education in occupational therapy, work experience in health care and personal knowledge gained from meeting specialized learning needs in her own family, she knew her energies should remain there. Margaret's knowledge of the medical system and sense of gratitude toward the health care system inspired her dedication to the hospital sector.

Education, in John's mind, is the cornerstone to career success, not to mention the basis of becoming a good engineer. Before U of T drew so much time and attention, particularly of John's, he and Margaret directed large gifts to organizations including the Princess Margaret Hospital, the National Ballet of Canada and the Canadian Opera Company.

While vacationing in the Bahamas, John happened to read an article that discussed the impact of the "double cohort" year and the potential shortage of university spaces for thousands of high school graduates in Ontario. Having been a student himself during the post-war shortage, he wondered how the University of Toronto's engineering faculty would meet the demand this time. The idea of turning away talented, aspiring engineers of tomorrow because there were too few places was one he could not support. He knew he would increase his assistance at that point. Dean Michael Charles happened to be spearheading a campaign for the Faculty of Applied Science and Engineering when John contacted him – a call that generated great activity. In 1998, he and William "Bill" Daniel (president and CEO of Shell Canada) co-chaired what would be the largest fundraising campaign in the history of U of T's Faculty of Applied Science and Engineering. This endeavour was not to be a hobby for John; it was his next mission, one that he pursued with the same limitless zeal that he would have given to a hinterland construction contract.

As Michael Charles now says of John, "A characteristic that came through is that he is forthright and persevering. I mean, once he grabs hold of something, you've got to have pretty good arguments to unseat his point of view. He is very strong-minded, in terms of what he wants to do, how he wants to do it and getting it done."

The idea of an Advisory Group had been percolating in the back of Dean Charles' mind for many years before his appointment as dean. He asked John to chair the committee, composed of engineer alumni and business executive members. John made it clear that he intended to be an active project manager with a very clear goal in mind, not a figurehead. His attention would be on programming and enhancing U of T.

Endowing Chairs was just one of the many further ways that John ultimately contributed greatly to the university.

John and Margaret with daughter, Susan, and her children, Scott and Katie, at Disney World.

JOHN EDWARD BAHEN
5T4 CIVIL ENGINEERING

Engineer Bahen was personally involved in some of the highest profile heavy construction projects in Canada during his 30 years of service (1964-1994) with Peter Kiewit Company Limited. He was President and CEO for the 1968-1994 period and was a member of the Board of the parent U.S. Company for 16 years.

Starting as a Project Manager, he was responsible for the construction of some 40 percent of Toronto's subway system, two sections of Montreal's Metro, and the elevated sections of Vancouver's Skytrain. Extensive road and dam construction throughout Alberta came under his overview, as did sections of the new Welland Canal.

He also carried responsibility for eleven projects related to the James Bay Hydro Facility, plus the concrete projection platform for the vast off-shore Hibernia oil drilling rig which came into service in 1997, adding to Canada's wealth and prestige.

He has played an active and effective role in support of human welfare organizations. The Sheila Morrison School for Learning Disabled Children and the Princess Margaret Hospital have received the benefit of his leadership. Cultural organizations such as the National Ballet of Canada, and the Canadian Opera Company were also the recipients of his support.

Since his retirement from Kiewit, Engineer Bahen has given outstanding support to the Faculty of Applied Science and Engineering with the establishment of the Bahen-Tanenbaum Chair in Civil Engineering, his Chairmanship of the Dean's Advisory Board and Co-Charmanship of the Fundraising Campaign Cabinet for the Engineering Faculty.

He has helped to build Canada, while providing leadership to the medical and cultural arts, and his Alma Mater.

Page from the Awards Banquet program when John received the Engineering Alumni Medal from U of T's Engineering Alumni Association.

21

PUBLIC RECOGNITION

IT was announced that John was to be awarded the Engineering Alumni Medal in October 1998. In a letter from Robert M. MacCallum, then president of the Alumni Association, John was advised that the medal is the highest honour the university can confer on one of its engineering graduates for achievement in their career. One of the goals behind the creation of the Engineering Alumni Medal was to foster connections between the Faculty of Applied Science and Engineering and its alumni. The more specific purpose for the medal was to improve awareness of the engineering profession by honouring distinguished alumni for their achievement. Success, however one measures it, is the common thread among the past medal honourees. They are outstanding role models for engineering students and representatives of fellow alumni.

The honours for John and Margaret continued. In 1996, John had received the Arbor Award for volunteerism on U of T's behalf, as did Margaret in 1998. As well, the gold medal award in the Occupational Therapy program was later established in Margaret's name by U of T president Robert Prichard.

The culminating and most humbling of John's recognitions by a grateful alma mater was the bestowal on him of an Honorary Doctorate in Engineering in 1999. The attention was surprising to the Bahens, because their active involvement in the university's growth was so personally gratifying, especially for John, who was enjoying himself immensely.

John receiving an Honorary Doctorate in Engineering from U of T in 1999. Applauding him are the university's chancellor Henry (Hal) Jackman (behind John) and president Rob Prichard.

Former dean Michael Charles delivered the Citation (see Appendix 2) on June 9, 1999, introducing John at the ceremony. He provided a summary of John's project involvement and his loyalty to the University of Toronto and philanthropic efforts, noting he is "one of Canada's greatest philanthropists and builders. His vision and generosity have been the driving force behind both his professional accomplishment and his remarkable volunteer efforts." Charles reminded those in attendance of the extent of John's reach across Canada on noteworthy and memorable projects. He concluded that "it is safe to say that there are few Canadians alive today whose lives have not been improved in some way by John Bahen's achievements in the construction industry."

In John's mind, receiving the honorary doctorate remains the ultimate achievement of his life. He realizes now the irony that, though he was an impatient, average student himself, at the end of his career he is an advocate for the necessity of a "book education" in pursuing engineering or heavy construction. The Honorary Doctorate in Engineering was rarely given, and in his characteristic way of injecting humour into everything, he said he felt more humbled than most would, "because I was not the best student by anyone's measure." John feels that his acceptance speech for the degree was the best he ever gave (see Appendix 3). Students at that 1999 convocation crowded the platform after giving him a standing ovation to shake his hand and offer a casual "thanks." Fifty-six years since his own post-graduate career began, he was more at home with a gathering of like-minded, eager new engineers than almost anywhere else he could imagine.

"They all spoke my language, without saying a word," he explains. "Being an engineer is about harnessing a relentless curiosity. We all have it."

Just when he could never have imagined a more prestigious recognition, he was advised that he was being awarded the Order of Canada. On May 31, 2001, then governor general Adrienne Clarkson invested him

John receiving the Order of Canada in 2001 from Governor General Adrienne Clarkson.

as a Member of the Order of Canada. Details of his status are noted thusly on the website of the governor general of Canada:

John Edward Bahen, C. M., D.Eng.

A highly successful construction industry executive, he directed landmark projects such as the Skytrain light rail system in Vancouver. A pre-eminent University of Toronto volunteer and benefactor, he co-established two Chairs in civil engineering, endowed the Michael Bahen Chair in epilepsy research and has contributed towards the establishment of a new centre of information technology. The National Ballet of Canada and the Canadian Opera Company have also benefitted from his generosity.

And then, on October 8, 2002, there stood John at the podium for the official opening of the Bahen Centre for Information Technology – having been surprised at the outset to learn that it was to be named after him. A quote by John and Margaret in the Winter 2003 edition of *Skule Matters* says: "We are thrilled to help facilitate world-class IT research, so that new discoveries that will have a positive impact on society can be made right here at the University of Toronto."

The Bahen Centre responds to a tangible need for places. It adds the lecture halls, labs, meeting rooms, offices and research facilities required to meet the dual needs of enrolment growth and the school's commitment to remain state-of-the-art in information technology.

The Centre is an architectural symbol of modernity and advanced technology, supplying spaces for students where multiple disciplines share teaching and research space. But the building is also a product of the Diamond Schmitt Architects' vision and sense of responsibility. Among Diamond Schmitt's own philosophical requirements of the project was that it address the street and that it, within the community in which it is located, use resources responsibly and not simply be worthy of attention.

PUBLIC RECOGNITION

The creative and efficient use of resources to achieve maximum quality were principles John lived by. It seems fitting that a building that would ultimately would bear his name was created with similar principles in mind. Drawing on his years as a U of T student, one particular design element that John requested for the large lecture halls was to have an entrance upstairs and an exit at the lower level. He hated the crowding caused by those trying to get out as others were arriving.

The Vertex in Surrey, England, which the Bahens bought in 1996.

22

LIFE CHANGES

By the late 1990s, John and Margaret were enjoying the level of comfort that success had afforded them. While they kept their house on Abaco, they sold their house in Toronto at 7 York Ridge Road and moved to England. At the start of 1996, they bought The Vertex, an architectural gem high on a ridge in Weybridge, Surrey, about twenty-six miles from Windsor Castle. Their choice of location was partly due to John's natural pull to England, his mother having been English, as well as his and Margaret's love of Europe. The Vertex was their home base from which they could indulge in their love of golf and boats by cruising in their custom-built *Argyll*, driving and golfing their way around Europe with flexible itineraries.

John ordered his life in his small, green, leather-bound notebook, with gold numbers indicating the year. The little book stayed in his shirt or shorts pocket, and in it, in pencil, he recorded meetings, appointments, flight times and places where he and Margaret would be staying while travelling. The first week of September 2001 noted planned visits to Inverness, Dornoch, Edinburgh and York. For September 7, he wrote Cambridge, followed by arrival back at The Vertex the following day. He noted a reminder for later in the month of needing an electrocardiogram appointment before heading to Belfast for golf. He also jotted down that, in late September, he would be at the Ryder Cup before finishing out the month in London.

However, control over his time and his life was taken from him on September 8, 2001. The Bahens were visiting their good friends Stuart and Joan Lee in Cambridge, where Stuart was a visiting dean at the university. While they were enjoying a light lunch, Margaret noticed John's left arm slide off the table where it had been resting at the edge of his plate. Looking at him, she saw one side of his face, at the mouth, drooping down. Realizing that he John was about to fall, she leapt from her chair, instantly aware that he was having a stroke.

An ambulance arrived within twelve minutes and took John to the Cambridge Hospital's Lewin Stroke Unit, where a CT scan revealed a massive blood clot in his brain. While the stroke's damaging activity continued within John, Margaret fought tirelessly to get access to an experimental clot-busting drug that the intern who was working closely with the Bahens mentioned was being tested by the hospital in clinical trials. However, she was exasperated to learn that, because the stroke had occurred on a Saturday rather than a weekday, it was impossible to access the medication. In a desperate frenzy, she thought in frustration of all the things she could do in any other situation to get John what he needed. They had the resources and knew the best doctors, and she had intimate knowledge of what the extent of this stroke might be without the aid of the latest drugs.

Three hours later, she quietly had to accept that every option had been exhausted when the team of doctors and nurses pushed the only blood thinners available through John's intravenous line. She knew, the intern knew – and they still believe today – that, had access to the trial drug been open to them, the extent of damage from the stroke would have been lessened.

Stuart and Susan flew overseas immediately to be with their parents. They departed just in advance of the tragic World Trade Center events of September 11, but consequently found that, due to airport closure and airspace monitoring over the days and weeks that followed, they could not get home as planned. While the world reeled in reaction to the attacks on America, the Bahen family faced John's stroke first with optimism and eventually with a realistic and proactive attitude not unlike the personality of the man himself. Some friends, who had been in Ireland when they got the news, arrived within a couple of days to support Margaret. Doctors told Margaret that John would never improve beyond what he was when the stroke ended. However, she was completely unwilling to accept such a hopeless prognosis at such an early stage. A call to Sylvia Lister, a friend whose husband, Noel, had recently suffered a stroke, convinced her to transfer John to the Wellington Hospital in St. John's Wood in London.

She and the children refused to believe that he would do any less than return to a high quality of life, not to mention that he would sit up and even walk. Wellington Hospital was terribly expensive, and it meant that for the eight and a half months he was there, Margaret lived in a hotel down the street in order to participate in his care. Once he was settled, the family felt strongly that if he was going to improve, Wellington would be where it would happen.

The Bahens had been planning to move back to Canada from England and had built a house on a property called Foxley Green, in King, Ontario. Some modifications were made to the home's architectural design to minimize level changes and incorporate elevators for John's

Cory Boyd presents John, surrounded by family and friends, with a jersey from John's beloved Toronto Argonauts.

ease of movement. For John, this was the beginning of the next project. Always optimistic and insistent that something good would come from every experience, he relied on his confidence in himself and the love and loyalty of the people around him to begin and continue rehabilitation. His humour continued to shine through, even throughout rehab, such as when he was being taught how to tie his shoes. Recognizing his situation – that he was in a wheelchair with just one functioning arm and with full-time care – he looked the therapist straight in the eye and said, "Do you really think I'm going to be tying my own shoes?"

John's biggest fear was losing his ability to enjoy the family that he treasured and the life he had worked for. His best friend, Denis Evans, knew that the unrelenting Bahen-style would re-emerge. "He was that guy that, if he got knocked down, was the first to get up." In a single-minded fashion, the entire family and the Bahens' closest friends embraced the commitment that he would continue to live as closely as possible the way he had lived pre-stroke. Some paralysis and memory difficulties slow him down, but he continues to attend some special events and rarely misses a football game. He still spends winters in the Caribbean on their 153-foot wheelchair-accessible motor yacht, which they bought in the U.S. and named the *Argyll*, after a yacht they had had built years before. He dotes on his five grandchildren, and in fact, Foxley Green was the perfect setting for their granddaughter Katie Chant's wedding, in 2010.

One story John enjoys telling is about a 2008 trip on the *Argyll* to Nova Scotia. He and the boat crew, all with the name "Argyll" on

The wedding of the Bahens' granddaughter Katie Chant, held at Foxley Green in 2010. Back row: Stuart and Kate Bahen (left), the groom Joe Armitage, Katie (Chant) Armitage, Susan (Bahen) Chant, Ted Chant, Jack Bahen, Scott Chant, Susie Bahen. Front row: Mikey Bahen (left), Margaret, John.
(Photo: Heartline Photography)

their shirts, were sitting outside a Tim Hortons in Halifax. They were astounded to learn that the restaurant had run out of donuts.

"My goodness," exclaimed John rather loudly, "if there are no donuts, it can't be Tim Hortons!"

Eventually, a downpour sent them scurrying back to the boat. (John describes his speed saying: "I guess my chair went faster with whoever was running on my behalf.") About twenty minutes later, a gentleman who had been sitting at the next table, having noticed the names on their shirts, arrived at the boat carrying a huge box of donuts. It turns out he was the president of Tim Hortons.

The *Argyll* was also the scene of a truly great celebration: John and

LIFE CHANGES

Margaret's sixtieth wedding anniversary, in August 2013, an occasion that brought the whole family together in Nantucket.

Today, John thrives on the occasions when Foxley Green is busy with people and the entire family is present. He's also just as content at his favourite place, their cottage overlooking the waters of Georgian Bay.

When John Bahen reflects on his life, he says simply, "It's been a happy life." When he ponders his career, his words echo the amazement of the boy who wanted to build things and grew into a man who accomplished more than he ever could have imagined:

"We built some pretty amazing things, didn't we?"

The first Argyll.

Experience fine living at sea...

ARGYLL

Commemorative coaster made by the staff of the Argyll for John and Margaret's sixtieth wedding anniversary.

Family celebration of John and Margaret's sixtieth wedding anniversary on the Argyll in Nantucket, August 2013.
(Photo: Claudia Kronenberg)

APPENDIX 1

Speech by John Bahen at the Opening of the Bahen Centre for Information Technology, University of Toronto, October 8, 2002

President Birgeneau, honoured guests, alumni, faculty, staff and most importantly, current students of the University of Toronto – good afternoon. Thank you for giving me the opportunity to talk to you today.

The good news is, clearly I'm speaking from the bottom of the batting order on the list of speakers. For those of you who know me, this is a position I enjoy occupying. I'm used to having the last word.

Now the better news! Today I am representing the individual donors who participated in the building of this centre. There is good reason for you to be appreciative of my being selected to speak on your behalf. Since my stroke a year ago, I cannot stand for very long anymore … so my speech will be short.

It is a very great honour for me to be here on behalf of the individuals and corporations who have contributed to this building. It is true that the Bahen Centre for Information Technology would not have been possible without the support of the provincial and federal governments. However, it was the generosity of the many private sector partners that was really the catalyst to making this project happen. I thank all of you for your vision, support and encouragement. We are tremendously proud to open this building during the 175th anniversary of the University of Toronto.

Information Technology has had a significant impact on business in virtually all sectors of our economy, and it will remain a key element of our country's development for the foreseeable future. In fact, the future of our country will likely walk the halls of this building some day, maybe even tomorrow. That makes me very proud.

I have been asked several times why I became involved in this project. Let me quickly tell you the story. Some five or six years ago, Margaret and I were enjoying a vacation, and I had some reading with me produced by the Province showing the demographics of how many high school students would be available for the next entrance to university. Almost 5,700 high school graduates would be shortchanged by our system – there was no place for them to continue their studies in Ontario. I reacted very sadly to that fact – which has proven to be a very expensive reaction for me, by the way. Both Margaret and I recognized we did not face this kind of constraint when we finished high school, and we realized the tremendous need that existed. We agreed that we could assist, and we could not think of a better way to contribute than making this IT building happen.

I would also like to thank two groups that are very important to me personally. Recognition of their outstanding performance in the delivery of this project should not be overlooked, but often is: the contractor, PCL, and the architect, Donald Schmitt. You and your teams are a credit to our industry. Thank you.

In closing, I have a strong message for our alumni. As you all are aware, *Maclean's* magazine annually researches specific comparative facts about Canadian universities. They compare us to our competition … do not think for a moment we are not in competition with the other universities – we are! One of the significant factors *Maclean's* uses to evaluate and compare us with others is the support the university receives from its alumni. The University of Toronto consistently rates number one. We need to keep it this way.

I have had, and continue to have, a lot of fun here. I want to leave with you a clear message for the alumni, both those past and those future. Take the time to revisit your university after you graduate, and recognize that your university needs your help and always will. Get on with the idea of giving.

On behalf of my family and all the other donors, thank you for making this day happen.

APPENDIX 2

JOHN BAHEN CITATION
UPON HIS BEING GRANTED AN HONORARY
DOCTORATE OF ENGINEERING DEGREE,
UNIVERSITY OF TORONTO, JUNE 9, 1999

READ BY MICHAEL E. CHARLES

Chancellor Jackman, it is now my great pleasure and honour to present John Bahen to you for the conferral of the degree of Doctor of Engineering, *honoris causa*. I have enjoyed the privilege of knowing John Bahen throughout my entire term as Dean of the Faculty of Applied Science and Engineering. During this time, John Bahen has been one of the Faculty's most dedicated alumni, conscientious volunteers, outstanding benefactors and best friends.

After graduating in 1954 with a BASc degree in Civil Engineering (as indeed, many present are about to do today), John Bahen began his illustrious career in the Canadian construction industry. Throughout thirty years of service with Peter Kiewit Sons, he was personally involved in some of the highest-profile construction projects in Canadian history.

Starting as a Project Manager with Peter Kiewit Sons and eventually rising to the position of President and CEO, John Bahen was responsible for the construction of approximately 40 per cent of Toronto's subway system, two sections of Montreal's Metro system and the elevated sections of Vancouver's Skytrain. During his career, John also oversaw extensive road and dam construction throughout Alberta, as well as construction of sections of the new Welland Canal, eleven projects of the James Bay Hydro Facility and the concrete production platform for the vast off-shore Hibernia oil-drilling rig. While overseeing the work of the V.K. Mason Construction Company, John was involved in the construction of such Toronto landmarks as the Hazelton Lanes complex and the CIBC Commerce Court.

All of these are massive endeavours, and it is unnecessary to emphasize the complexity, or importance, of these projects, but John Bahen earned the reputation of completing the projects ahead of schedule and with unparalleled quality. It is safe to say that there are few Canadians alive today whose lives have not been improved in some way by John Bahen's achievements in the construction industry. Although his work often took him from home and family for long periods of time, it was with his wife's help and support that John was able to make this sacrifice and complete these vital infrastructure projects. It is a pleasure to acknowledge that Margaret Bahen is with us today.

John Bahen's service to the Faculty of Applied Science and Engineering is highlighted by his tireless work as a volunteer. For several years, John has served as Chair of the Dean's Advisory Board, a group of leaders from industry and the engineering profession who offer their expertise in helping to chart future directions for the Faculty.

When the University of Toronto launched its current campaign to raise funds for student support, endowed Chairs, research facilities and key capital projects, John Bahen accepted the volunteer position as Co-Chair of the Campaign for the Faculty of Applied Science and Engineering. John has consistently proven himself to be a great leader for the campaign, helping the Faculty to achieve its vision by reaching out to the broader community, building bridges, making friends and enthusiastically enlisting support from his alma mater.

In 1995, John Bahen was responsible for co-establishing, with fellow alumnus Joey Tanenbaum, two endowed Chairs in civil engineering.

These were the first endowed Chairs ever created within the Faculty, and they provided our campaign with great early momentum. The first incumbents of these Chairs are Professors Richard Soberman and Michael Collins, two of Canada's leading educators and researchers in transportation and structural engineering, both of whom are here today to hood John on this important occasion.

The impetus provided by establishing the first two Chairs cannot be underestimated; in the meantime, a total of sixteen Chairs are now in place or to be announced shortly. Chairs provide a means to add to the Faculty complement, thereby improving the student/Faculty ratio and extending the range of expertise available to our students.

More recently, John and Margaret Bahen together donated $2 million to establish the Michael Bahen Chair in Epilepsy Research at the University of Toronto in memory of their son. The first incumbent of the Chair, Dr. Mac Burnham, will work to generate new insights into the causes, effects and treatments of epilepsy, and will attempt to de-stigmatize the perception of epilepsy in society.

Even more recently and earlier this year, John and Margaret Bahen have donated $6 million toward the establishment of a new building to be known as the Centre for Information Technology at the University of Toronto. The Centre will provide the facilities necessary for the Faculty to double its enrolment in high-demand engineering programs based on information technology, responding to Canada's urgent need for qualified IT practitioners. It is safe to say that this initiative would not be on the fast track that it now is without the Bahens' vision, sense of urgency and generosity.

John Bahen's dedication to the University of Toronto has been overwhelming. Few people are more committed than John to helping our Faculty fulfill its goal of becoming one of the world's leading engineering schools.

John and Margaret have also played an active philanthropic role

APPENDIX 2

outside the university, supporting human welfare organizations such as the Sheila Morrison School for Learning Disabled Children and the Princess Margaret Hospital, as well as cultural organizations, including the National Ballet of Canada and the Canadian Opera Company.

John Bahen is one of Canada's greatest philanthropic leaders and builders. In fact, I think that the only project that he has ever failed to complete properly or on schedule is his own retirement, which I'm afraid is destined to remain on hold for quite a while. He exemplifies all the finest qualities of graduates of the Faculty of Applied Science and Engineering and the University of Toronto: foresight, dedication, leadership integrity and perseverance, all coupled with a warm heart.

I am so proud to know John and Margaret Bahen, both absolutely outstanding people.

Mr. Chancellor, on behalf of the Governing Council, I ask you to confer the degree of Doctor of Engineering, *honoris causa*, upon John Bahen.

APPENDIX 3

An Address Presented to the University of Toronto Graduating Class of Engineers Receiving their Bachelor of Applied Science Degree, June 9, 1999

by John Bahen
A former graduate and retired President of Peter Kiewit Sons Co. Ltd., and currently Chairman, Dean's Advisory Board, Faculty of Applied Science and Engineering

Developing Yourself

Chancellor Jackman, President Prichard, Chairman Cecil-Cockwell, Dean Charles, distinguished graduates, families, friends and especially you engineers.

I would like to begin by thanking the Governing Council of the University of Toronto for bestowing upon me the outstanding honour and privilege of addressing you this afternoon. Thank you.

Throughout my career, I have always been amazed at how little I knew at the time of my own graduation. As I look out at today's graduating students, I realize that university is a solid foundation for a life of learning, and that my own experience at the U of T gave me a great start in the process of developing myself.

I am so pleased today to have received this honour that I feel obliged and vigorously anxious to share with you some experiences that I hope will assist in developing yourselves in the years to come.

APPENDIX 3

Let me begin by recognizing the importance, at this pivotal moment in your lives, of developing personal goals. I believe that a worthy goal for all engineers is to develop the people skills and business skills that will complement your engineering education. Today, the University of Toronto is attracting and educating leaders of tomorrow, and you will always be identified as part of that group. You will soon realize that engineering is a challenging profession. I have full confidence that you, as bright, ambitious people, will seek out the opportunities and not back down from the challenge ahead.

For the first few years after graduation, your challenge will likely be to make that crucial transition from student to engineer. You will be preoccupied with learning your trade, learning what it means to be an engineer in today's world. After all, an engineering degree won't make you an engineer any more than a medical degree would instantly make you a doctor! It takes practical experience to become an engineer.

Today, as you receive your diplomas, you can say that you have fulfilled the necessary prerequisite, you have completed your formal engineering education. And as you think about the significance of this day, you should take pride in the fact that you have chosen a difficult academic program, and you have succeeded. Take a while to enjoy that success.

After the first few learning years of your career, the greatest challenges that you will face will not likely be engineering related. They will most likely be communications problems, people problems and business problems. Not bad problems to have, so long as you are prepared to handle problems, a very significant first goal for you.

A second goal, which may be easy to forget today, is to maintain good health! Forget about smoking, Control your "40-beer" intake! You know better than I that vigorous daily exercise is necessary. Developing yourself means developing the physical vigour to be aggressive and well balanced.

APPENDIX 3

Please, please approach all future challenges with a sense of responsible optimism. Whatever you want to do, you can do it, so go do it! Be humble, but believe in yourself. Don't ever give up when you know that you are right – don't allow yourself to get distressed when all about you are questioning your direction. Always remember, the fact that you are here today means that you are not ordinary people, you are achievers! Despite your humility, as engineers you have the potential to create, the discipline to work long and hard and the talent to arrive at the right conclusions.

I have a real life experience to share with you, in order to illustrate the rewards of hard work and perseverance. The Hibernia oil platform that came into production twenty-one months ago is the climax of my story.

Let me take you back to the year 1990. My company, Peter Kiewit Sons, was not invited to bid on this major construction project despite the fact that we were the major concrete structures contractor in Canada and the U.S., and especially in eastern Canada. The Mobil Oil Company from Houston didn't want us, but we wanted their job! The problem was that we had no experience working in the North Sea off Norway.

We finally bid on Hibernia with an alternative price from another fabrication site. During the tendering period, my chief engineer and I spent four or five days near Clarenceville, Newfoundland, and late Saturday night I was reading the Newfoundland road map when I spotted a town called Sunnyside. The name caught my eye because it seemed to be a misnomer for any town in Newfoundland, having experienced their weather while on construction in Goose Bay, Labrador, for three years. The next day, Sunday, we found a fisherman who took us in his boat and showed us around Sunnyside.

Our proposal to Mobil included this alternate site. Even though Mobil had previously dumped us because of our lack of North Sea experience, they decided to recall the bids and include our site at Sunnyside as the place to set up the fabrication yard.

APPENDIX 3

This recall of bids resulted in our being the second bidder. Why give up when you're second? After all, you have nothing more to lose. Despite our efforts, Mobil decided to award the $1.2 billion job to a five-way Paris, France, venture. Twenty months after they started construction, we moved in, at the request of Mobil, to take over and complete the job. Our company knew how to build this kind of job, and control, train and motivate the work force. This billion-dollar job was very successful for the Kiewit Company.

Now what lesson did I learn from this "war story" that I can pass on to you today?

- Be humble and communicative.
- Be vigorous in your efforts to achieve your goals.
- Retain personal willpower with some patience.
- If you know you are right, never give up.
- The most important thing to believe in is yourself.

What unique accomplishments do you seek? As future leaders, you have a responsibility – be innovative – have an aim, have goals that excite you and never give up in the pursuit of those goals. Go to work every day with a feeling of challenge and excitement. Make life worthwhile. In no way should learning stop upon graduation. Your education can now begin. Significant changes and opportunities will continue to reveal themselves to you. There is an important relationship between change and opportunity. Change is difficult, but you must be willing to accept this opportunity.

When I reflect on the unlimited world of Information Technology, the explosion of the Internet and the fact that "go ahead" companies design every aspect of their equipment to be continuously upgraded, I realize that the key to success in today's world is to embrace and harness the benefits of change. You, too, have to meet the challenge of continuously upgrading yourselves. Change is constant.

With the hockey season about to come to a close, NHL owners are teaching us all a lesson about change as they look for ways to make the game more exciting. They are considering widening the goalposts and eliminating the red lines; very significant changes indeed! And even if these changes are not successful, they will learn from the experience, teaching us that we can all learn more from setbacks than from success.

How do we learn from our setbacks? Remember the details and continuously ask yourself how you could have recognized and responded to the problem sooner, If you experience a corporate setback, participate vigorously in the corrective action, face up to the problems, communicate them. People will remember you not only by your successes but also by how you resolved your setbacks.

I believe that it is important to develop your leadership qualities, since many of you will be leaders and managers in your careers. The important personal qualities of a leader or a successful manager are basic honesty and a willingness to work hard. These qualities are indispensable in gaining the support of employees, associates and clients. A leader must be the type of person who inspires by example, so that others can look to him or her for direction and guidance. A person with these qualities will, gradually, earn a position of leadership, and if that person ever fails to merit that position, it will pass to someone else.

As I look back on the people I consider inspirational leaders, I find that they were all individuals who liked and understood people; who had patience and understanding; who set objectives and, where appropriate, delegated authority; who made the work interesting; and who set the highest standards, both at work and in their personal lives. These leaders challenged and motivated those of us working for them so that we were always able to do more and aspire to greater achievements.

APPENDIX 3

Before I conclude, let me just leave you with a few characteristics that may help you develop yourself into a leader of tomorrow.

Personal discipline and order in your life are important. Professor Michael Collins once wrote in a report that, "It is far too easy to let urgent displace important." I agree, and I believe that discipline and good judgement will overcome this problem.

Develop empathy for others: appreciate the value of someone else's viewpoint.

Develop the ability to listen: good managers know how to listen and seek advice from others. Ironically, when you rise to a top management position and you have more say in important matters, this is when you begin to say less and listen more.

Develop good writing skills: in your job, you will find that you will do a lot of it, and you will find that the better your ideas are communicated in writing, the more people will respond.

Learn to work with people and manage people: no one in the real world works in a vacuum. Nearly everything we do as engineers relates to people in some way. Be comfortable with people.

Never stop learning: make sure that you, and those around you, always make self-improvement a priority.

Finally, possess a high degree of optimism and happiness about your career status, your job and the world in which you live. Challenge is exciting. And the world is full of challenge. Having a sense of urgency increases the challenge.

So there it is – my advice on developing yourselves into the leaders of tomorrow – given to you by a seventy-year-old who believes that he can say it as it is because he has lived it. So go out and build a better Canada for the next century!

Thank you, and I wish you the very best in your exciting life.

REMEMBERING
MARGARET AND JOHN BAHEN

Margaret and John Bahen (CivE 5T4).
Were both University of Toronto alumni, their visionary philanthropy is seeding breakthroughs in medicine, engineering, math and computer science.

The University of Toronto has lost two remarkable alumni and supporters. **Margaret and John Bahen** (CivE 5T4) both passed away in November, within days of one another. The couple, who met at U of T and raised three children together, leave behind many friends and family members, as well as a strong legacy at their alma mater.

The Bahens gave generously to the Faculties of Medicine and Applied Science & Engineering. Their commitment to advancing medical research and scholarship, and their lasting contributions to the campus through their support of the Bahen Centre for Information Technology, leave an indelible mark.

"The University is proud to count John and Margaret among our most distinguished alumni and champions," says U of T President Meric Gertler. "We greatly appreciate their dedication to advancing excellence in research and scholarship across disciplines, and their visionary understanding of how shared space in state-of-the-art facilities fosters innovation and collaboration."

John and Margaret's legacy of philanthropy to U of T Engineering is embodied in the Bahen Centre for Information Technology. Completed in 2002, the building houses advanced research and teaching facilities, which support faculty, staff and students in the Faculty of Applied Science & Engineering, the Department of Computer Science and the Department of Mathematics.

"It is impossible to overstate the impact of John and Margaret's generosity," says Cristina Amon, Dean of the Faculty of Applied Science & Engineering. "Their support of engineering research and the visionary design of the Bahen Centre as a focal point for collaborative information-technology research will be felt by U of T and Canadians for generations to come."

A 1954 graduate of the Department of Civil Engineering, John Bahen began his career at McNamara Construction, specializing in large-scale projects such as hydroelectric dams and major highways. In 1980, he was appointed President of Peter Kiewit Sons' Co. Ltd., where he directed landmark projects such as the SkyTrain light rail system in Vancouver. Upon his retirement in 1994, he and classmate Joey Tanenbaum co-established the Bahen/Tanenbaum Chairs in Civil Engineering, which focus on applications of structural engineering.

Professor **Michael Collins** (CivE) held one of the endowed chairs for 19 years. "While engineering professors should conduct research, teach and provide service to the community and the profession, it is my belief that teaching students the basic principles of the art of engineering is the central role," says Collins. "The Bahen/Tanenbaum funds have been a significant factor in enabling me to greatly increase the number of students I teach while still maintaining very high standards."

"In making a generous commitment of time and financial support to the Bahen Centre for Information Technology at a crucial time in the Faculty's development, John and Margaret Bahen had a major and lasting impact on the Faculty's quest to be counted among the world's leading engineering schools," said **Michael Charles**, former Dean of the Faculty of Applied Science & Engineering. "My wife, Barbara, and I are grateful that we had the opportunity to know and enjoy the friendship of Margaret and John Bahen, and to witness their admirable loyalty to the University."

At the Faculty of Medicine, the Bahens' support is making a major impact on health care. "John and Margaret Bahen have left a great legacy in medical research and scholarship," says Trevor Young, Dean of the Faculty of Medicine. "Their support continues to be felt through promising work in epilepsy and occupational therapy."

In memory of their late son Michael, Margaret and John Bahen created the Michael Bahen Chair in Epilepsy Research. Chair-holder Dr. Berge Minassian has discovered genetic mutations underlying several types of epilepsies, including the severest type – Lafora disease. These discoveries have led to the identification of a potential drug to counter the disease, which will soon be undergoing clinical trials. The Bahens also contributed generously to the Epilepsy Research Fund within the Clinician Scientist Training Program in the Department of Paediatrics.

"The Bahens have brought us to the threshold of overcoming severe epilepsy," says Minassian. "I am extremely grateful for their support, which has been crucial to this research and continues to push this area of medicine forward."

Following her graduation with a diploma in Occupational Therapy in 1952, Margaret Bahen worked at Sunnybrook Hospital's Veterans K Wing. She created the Pamela Cowie Gray Generosity of Spirit Award in the Department of Occupational Science and Occupational Therapy, to honour her lifelong friend from the program, and remained connected to the department and to U of T's rehabilitation sector.

"We are tremendously grateful for Margaret Bahen's support," says the department's Chair, Susan Rappolt. "As an alumna of the program, she understood the value of hope and engagement as a resource for health, and was dedicated to helping future students pursue research and professional development."

In 2001, then-University President Robert Prichard and his wife Ann Wilson established the endowed Margaret Bahen Gold Medal in Occupational Science and Occupational Therapy, a convocation award given annually to a student excelling in academic courses, clinical fieldwork and overall leadership.

"John and Margaret were strong champions of our faculties of Medicine and Engineering, demonstrating their belief in the ability of academic research and higher education to transform people's lives and build better societies," says David Palmer, who is Vice-President Advancement of U of T. "The Bahens were standard-bearers of philanthropy in this country and will be remembered for their thoughtful and generous vision."

JESSICA MACINNIS
SENIOR COMMUNICATIONS OFFICER
THE EDWARD S. ROGERS SR. DEPARTMENT OF ELECTRICAL & COMPUTER ENGINEERING

JOHN EDWARD BAHEN December 27, 1927 - November 18, 2016 Dad passed away on Friday at age 88, son of Katherine, (nee Simmonds) and George Bahen and brother of G. William Bahen (Anne). Loving husband of Margaret Edith (nee Campbell), best father to Stuart, Susan Chant (Ted) and Michael (1984), grandfather to Katie Armitage (Joseph), Scott Chant, Jack, Susie and Mike Bahen and great-grandfather to Hunter Armitage. John grew up in Toronto and graduated from Vaughan Road Collegiate and the University of Toronto in Civil Engineering. He was a natural builder and people person. He started his career with McNamara Construction and finished as President of Kiewit Canada in 1994. With his team he successfully built some of Canada's largest and most difficult civil works, including dams and powerhouses, among many other projects. He was always an optimist. He would chase opportunities long after others would have quit. He was a true leader and developer of people. John and Marg's philanthropy over the years has greatly enriched Canada's medical, academic and cultural communities. John was awarded an honorary Doctorate of Engineering from his alma mater and received the Order of Canada and The Queen's Diamond Jubilee Award. An avid boater, he never saw a boat he didn't want to buy. And at the helm on Georgian Bay, he never had enough time to get where he wanted, was oblivious to his wake and was too well known to the shoals. At home, he was the best husband, father, grandfather and great- grandfather we all could have imagined. He greeted everyone with his big warm smile and that twinkle in his eye. Dad was always generous and kind. Fifteen years ago, Dad suffered a devastating stroke. Despite his incapacity, he remained gracious, optimistic and never lost that twinkle in his eye. Mom poured her heart and soul into Dad's care and comfort all of their last years. Our thanks to Fely, James, Gerald and Eric who have gently and lovingly looked after Dad for many years. A memorial service will be held at 2 p.m. Friday, December 9th at All Saints Anglican Church, 12935 Keele St., King City. In lieu of flowers, consider a donation to the Margaret Bahen Hospice for York Region, c/o the Southlake Regional Health Centre Foundation.

PUBLISHED IN THE TORONTO STAR FROM NOV. 24 TO NOV. 27, 2016

MARGARET EDITH BAHEN (nee CAMPBELL) January 19, 1931 - November 21, 2016 Margaret passed away from complications from ALS at her home in King City at age 85. Daughter of James Alexander Campbell and Catherine Edith (Maclean) Campbell, predeceased by her sister Mora Zimmerman (Hugh) and brother Mac Campbell (Marina). Wife of 63 years to John Edward Bahen, mother to Stuart, Susan Chant (Ted) and predeceased by son Michael. Grandmother (Bushey or Bad Granny) to Jack, Susie and Mikey Bahen, Scott Chant and Katie (Chant) Armitage (Joseph) and great-grandmother to Hunter Armitage. Marg grew up in Toronto, spent her school years at St. Clements where she started many lifelong friendships. Marg was an Alpha Gamma Delta at University of Toronto and graduated in Occupational Therapy. After university, she worked at Sunnybrook Hospital's Veterans K Wing. Marg and John met at University and after marrying moved to various job sites before starting their family. Marg raised three children while John built infrastructure around the country. Summers were spent on Georgian Bay, and later in life, winters in the Abacos. Family was a priority to Marg and she orchestrated many fabulous adventures and created wonderful memories with her grandchildren. In retirement, Marg and John travelled extensively and enjoyed boating in Europe and the Caribbean. Marg was blessed with many friendships and loved spending time with them and her many courtesy nieces and nephews. Marg was an avid golfer and card player. She was a part owner of a gift shop in Unionville, before giving it up to travel with John. For the past 15 years, Marg has been dedicated to caring for John after his stroke and continuing to provide an adventurous and stimulating life for them both. Margaret supported and volunteered for many causes, including research in Occupational Therapy and Epilepsy at University of Toronto, North York General Gift Shop. She was especially dedicated to her fundraising for Southlake Regional Health Centre's Rehabilitation Centre, Cancer Centre, Diagnostic Imaging Department and the new Hospice. She was known for her kindness and generosity, fabulous style and wonderful sense of humor. Her hospitality, whimsical collections and her tiny dogs were all part of time spent with Marg. Mom was lucky to have so many wonderful friends and relatives. Mom had an energy and spirit that light up a room and she will be dearly missed. Family and Friends are invited to celebrate Margaret's life at All Saints Anglican Church, 12935 Keele St., King City, on Friday, December 9th at 2 p.m. The family would like to extend our thanks to all of Mom's care workers, especially Fely, Michelle, Helen, John and Jerome. In lieu of flowers, donations can be made in Marg's name to the Margaret Bahen Hospice for York Region, c/o the Southlake Regional Health Centre Foundation.

www.ingramcontent.com/pod-product-compliance
Lightning Source LLC
Chambersburg PA
CBHW040800240426
43673CB00015B/403